COME IN TO LEARN
AND GO OUT TO SERVE

Dr. Darsh Wasan

Published by the Dr. Darsh Wasan

Darien, IL – USA

The view and onions expressed in this book are the authors' own and the facts are as reported by them which have been verified to the extent possible, and the publishers are not in any way liable for the same.

ISBN 978-0-578-58768-4

ADVANCE PRAISE FOR THE BOOK

"Dr. Wasan changed my life for the better. Under his tutelage, I learned how to pursue excellence through rigor, unyielding attitude, thinking outside the box, and managing time and more importantly, how to communicate with academic and business stakeholders. His positive energy, excitement, and genuine interest in his students' growth are contagious and made IIT Chicago and its Department of Chemical Engineering a better place to learn, grow, and prepare for the future."

— Angsuman Dutta (M.S. CS '02, M.A.S. ChE '02)

"I was a graduate student in Darsh's research lab in the early eighties. To a young foreign student like me, Darsh was far more than a research advisor. He was my mentor, guide, and surrogate father. His passion for exploration and innovation is contagious. A master storyteller, he energizes everyone in his orbit with his ideas, insights, and eloquence. I am deeply grateful to Darsh for making me a better scientist, a better communicator, and most importantly, a better human being."

— Venugopal B. "Venu" Menon (Ph.D. ChE '86)

"Professor Darsh Wasan is the most energetic, hardworking and student-focused professor I have come across in my life."

— Dharam Punwani (M.S. ChE '67)

"Professor Wasan was not only a thesis advisor for us, but also a father figure. His support for his students continued well after their graduation."

— T. S. Ramakrishnan (Ph.D. ChE '85)

"As all his students will testify, Doc Wasan is one of the most knowledgeable, skilled, and demanding teachers in the field of chemical engineering. On a personal note, he is an incredibly kind, considerate, and humble human being. A lack of culinary skill is his only flaw, which is why we all celebrated when he brought his lovely wife, Usha, into our lives."

— Ravi Randhava (M.S. ChE '66, Ph.D. '70)

"I was privileged to have Professor Wasan as my Ph.D. advisor. His reputation spanned the continents. I came from India just to be able to work with him. Not only was he an accomplished academician, he was also a great communicator and a fantastic human being. His uncompromising standards for quality and hard work were a beacon for his students and colleagues. He personally touched the lives of everyone around him with his kindness. I am forever grateful to him for a being a father figure as I was navigating my early professional life."

— Seth Rao (Ph.D. ChE '00)

"I came to IIT in August of 1986. I had a difficult time adapting and had decided to withdraw from IIT. I was planning on heading back to India, when I met Dr. Wasan for the first time and told him about my decision. He listened with empathy and then proceeded to tell me that in the Gita, when Arjuna has doubts about his role in the ensuing battle, Krishna tells him to focus on the task not on the fruits of the labor related to the task. He asked me to take three core courses, work hard, and see him at the end of the semester. I ended up working in his lab and have never looked back. His role as a leader has had a lasting impression on me."

— Chandrashekar A. "Shekar" Shetty (M.S. ChE '89)

DEDICATION

To my Waheguru Ji for his blessings.

CONTENTS

ACKNOWLEDGEMENTS

Nearly two decades ago, my good friend Dinesh Shah of the University of Florida encouraged me to write my memoir; he even offered me his condo in St. Augustine Beach, Florida, so that I could write over a semester break, especially in the winter months.

I was too busy advising my many students and keeping my research going. Moreover, I was traveling abroad to establish offices in Europe and Asia to dramatically increase IIT's international visibility, recruit students, and develop relationships with numerous universities.

I had logged over one million miles in less than 20 years, but foreign travel eventually took a toll on my health. After I got sick in 2013 in Beijing, my overseas travel had to be curtailed. Only then, after completing 50 years of devoted service at IIT and being prompted by my family and Sam Pitroda, my longtime good friend, I began to give serious thought to writing my memoir.

I do not know how to type, but a year ago and with the assistance of Rahul Madhusudhan, a graduate student in the School of Applied Technology, I began typing the memoir with a stylus on my iPad. I then sent a rough draft of each completed chapter to Kelly Lee, who over the years had edited many of the scientific manuscripts that my students and I published in professional journals. She corrected my English, checked facts, and expanded the text to be more interesting.

I then sent the revised chapters to my former colleague Serope Kalpakjian, who retired from IIT in 2001 and now lives in Boise, Idaho, near his son and his family. Serope has written excellent engineering textbooks, widely used and translated; I have always admired his lucid writing style. He edited all the chapters, added subheadings, and essentially rewrote the epilogue; in my sincere opinion, he deserves to be the co-author of the memoir. I am indebted to him for spending a great deal of his time on this project.

Robin Hunt, an extraordinary copyeditor, edited the memoir, for which I am most grateful. I am most appreciative of Scott Benbrook‘s work on the book jacket design, photo production and recreation of illustrations and tables.

Many of my colleagues, including Tom Jacobius, Louise Hewitt, Ophir Trigalo, and Rollin Dix, read the initial drafts of some sections, did fact checking, and provided supplementary information, which I gladly included in the final version. Dharam Punwani, one of my former students, made several good suggestions, which I have incorporated into the text.

I wish to thank D. "Raja" Rajagopalan, Tim Zhang, and Vanita Misquita for their contributions, which I have included, verbatim, in the memoir. Mary Dawson, associate vice president of international affairs, and Tracy Allen, my executive assistant, helped me in several ways. A special thanks to Alex Nikolov, my coworker, for his assistance with some of the photographs included in the memoir.

I am grateful to Usha, my wife, and to Ajay and Kern, my sons, who assisted me in writing several sections of the manuscript and reminded me of several facts that I had forgotten.

Finally, I wish to acknowledge my Waheguru Ji, for His many blessings.

Chapter 1
Beginnings

"Come in to learn and go out to serve."

How could I ever have imagined that this old school motto would someday become the inspiration for my entire life! My life has been an incredible journey. It may be hard to believe that it all began during one of the most challenging times in India's rich history, going back five millennia: the British partitioning of India in 1947.

Early Days

I was born on July 15, 1938, in Sarai Saleh, near the city of Haripur Hazara, in the North- West Frontier Province of British India. It is a beautiful area, rich in natural resources. I did not know my true birthdate because none of us had a birth certificate; I learned about mine later, when I received my high school graduation certificate.

I was the fourth of eight children in a middle-class Hindu family. My parents named me Darshanlal Tilakchand Wasan, after my father, Tilakchand. He worked as a wholesale grain merchant; although not educated, he was intelligent and spiritual. He was very strict with me and my siblings. Although frugal, he gave 10 percent of his earnings to charitable causes and helped poor people with their weddings.

My mother was not educated; she was a homemaker and a religious, loving person. The most favorite of our family, she had excellent relationships with everybody. When she died, I remember well crying most of the day. I was 46 years old, living in the U.S., and so far away from home; it was one of the most disruptive days in my entire life. She was the rock of our family and her sacrifice to provide for me and my seven siblings was unworldly; she will always be remembered as the light of our lives. I believe that my parents' practice of the Sikh religion and spirituality had a deep influence in my adult life. I learned to be detached from worldly possessions. It taught me that happiness comes from serving others and not from things.

My family was well respected in our village, where we had lived for several generations. We owned a farm, where we grew corn and other vegetables; we also had a cow. We went to school regularly, but we still had plenty of time to play and, especially, to dream of our future. We didn't have any toys or games; instead, the older boys would look for long pieces of soft wood, and with much care, they would carve out very simple toys and give them to the younger children to play with.

The simplest was a stick, used in a game called *gulli danda* (meaning the game of tip cat in English; similar to the game of cricket), played by boys in India for more than 2,000 years. There are several variations of this game, but basically a player uses a large stick *(danda)* to hit a smaller stick *(gulli)* on the ground. He then attempts to hit the gulli as far as he can. Our dusty streets were the perfect place for the boys to congregate in anticipation of the launching of another gulli danda!

The Panchyat

My family had lived in Sarai Saleh for a long time, where my grandfather had been a member of the *Panchyat Raj* (meaning ruling assembly of five in English). These assemblies or councils consisted of five wise and respected elders from the immediate area. Each group operated as the local government, overseeing such matters as land distribution, disputes over trade issues, and various local legal matters. The council was comprised of two Hindus and three Muslims.

The head of our village government (the chief administrator) was Firoze Khan, a Muslim and a wise, kind man. My brother, Madanlal (we called him Madan), attended high school with Khan's son. Even though Madan was a little older than me, I was very close to him. He was always a great example for me to follow; this was particularly important when it came to his untiring devotion to learning.

Every day, he would walk three miles to go to school in Haripur. He graduated from high school at the top of his class, a testament to his dedication and drive. I did not truly realize how important Madan was to me until much later in my life.

The Partition

Living on a farm, we had our daily routines. In 1947, however, when I was nine years old, my childhood took an abrupt turn when the British began to leave India. The country was portioned into East Pakistan (now Bangladesh), West Pakistan (now Pakistan), and India. The peace that had existed between the different religious groups for generations was now shattered in a sudden, shocking eruption of bloody sectarian violence. It resulted in unprecedented mutual genocide and mass migration.

The Hindus and Sikhs migrated east, to India, and the Muslims migrated west, to Pakistan. Millions became refugees, a roaring wave of humanity that swept up my family, eventually depositing us in Bombay (now Mumbai).

My family had never lived on any other land but was now suddenly forced to abandon its home. We could not have traveled safely without the assistance of Firoze Khan. He was extremely concerned about the increasing reports of violence that were creeping closer to our village. Quickly assessing the situation, he realized that we might be in real danger. Quietly, Khan helped us arrange to leave our home by train overnight to escape a possible attack by Muslims.

This was a shocking new reality for our family. Although we had been part of a tiny community of a few hundred Hindu families, in a town with several thousand Muslim inhabitants, we had always enjoyed good relationships with our neighbors. Until the partition, many Hindus even had owned businesses and had employed Muslims without problems. This situation made our journey even more desperate and painful.

Refugees

Our escape in the middle of the night has been one of my darkest memories. I was filled with terror as we approached the Indian border, with vivid memories of the trains that had all been attacked by Muslims. Like most refugees, we arrived in Bombay with only the clothes on our backs. We were sheltered in a refugee camp in Koliwada, in the suburbs; we were one more tiny addition to the swarm of desperate, fleeing humanity.

Shortly after our arrival, Khan's son brought us news of our village. At the time, wecould not possibly have understood all of the implications of India's partition. My father was confident that we would be able to return to the only home my family had known. He held hope dear to his heart, but he was to be disappointed.

Life in a Refugee Camp

Our family of 10 was allotted a ration card, a single room that was about 300 square feet, and a small kitchen. We had no electricity and no adequate toilet facilities. At night, we slept mostly outside on the ground; during the summer rainy seasons, we slept inside. It would rain for three months, cooling us off from the intense heat but also thoroughly drenching everything.

Rainy seasons were no small trial for small boys. In the U.S., the dog might have eaten your homework, but the rains in India could literally consume it. Our homework had to be written in ink, and it would often be wrecked by the daily rain and high humidity. I remember well waking up early some mornings to completely redo my assignments before going to school.

The refugee camp had a *gurudwara*, a Sikh temple (meaning door to the Guru). Unlike other places of worship, anyone could go there, even those who did not have any particular faith. It was and continues to be a welcoming place for all people, and it came to have a deep meaning in my life. My parents used to attend every morning, but because I was busy with my schoolwork, I went only on Sundays and holidays.

My Schools

I had completed the third grade before the partition. As a child, I did not enjoy going to school because our teachers were very strict; they even punished those who were late to school. However, my academic life changed completely when I began fourth grade, at the new refugee school, called *Guru Nanak*.

When I was younger, all of our classes were taught in Urdu, but the Guru Nanak offered lessons in Hindi and included Basic English, Punjabi, and, in the eighth grade, Sanskrit. School days stretched from seven in the morning until noon. They only gave us five hours because the school was so packed with earnest students that they had to have two separate sessions, including one in the morning and another from noon until five in the afternoon. The teachers lived on the campus as well. I spent my formative years in this rich, eclectic, and colorful environment, in a nation that was beginning to change rapidly.

I suppose my first stint of leadership and a career in teaching was formed in the fifth grade, when I was given the privilege to give talks to our entire student body—a special assignment that the teachers gave me for whatever reason. I suppose they saw a teaching candidate in me, so I used to give mini lectures to the student assembly after the weekly morning prayers ended. Two years later in the seventh grade, we had class monitors, who basically were the leaders when the teachers weren't there. Perhaps you can guess that I, too, was chosen as a class monitor, which proved to be one of my first inspirations in learning to lead and work with other students.

Students in the U.S. may run or play sports in their P.E. classes, but we had a special assignment in ours: carrying bricks. I have fond memories of how we all pitched in to carry thousands of bricks, which eventually became a new school building. When completed, our principal, Lakshman Singh, had the following motto inscribed on the front of the building: “Come in to learn and go out to serve.”

Examinations

I may not have realized at the time that my education would eventually allow me to serve to the fullest extent of my capabilities. Still, however, I had an enormous hurdle to overcome before I could make a difference in the world—the dreaded state exams.

The Indian High School Graduation Examination was a statewide examination that students had two months to prepare for. It was an incredibly stressful time for the students and their families. I was committed to doing well, despite the fact that our living conditions were very poor. I used to grab any time that I could to study, even going so far as to read under the streetlights by the school. I spent hours and hours studying, not in a library but on the streets near my home. I remember well memorizing math theorems by writing the proofs on the dusty ground. One night I fell asleep out there, and my brother Madan woke me up. I have deep memories of those days studying; they were the foundation of my training and work habits to get the job done no matter what—for we had limited areas in which to study and limited resources, still it was up to us to find a way to get it done. Little did I know that those deep-rooted study habits by the gravel road would be the foundation of a lifetime of researching and learning.

College

The decade-plus of crunching the books and having a singular focus finally paid off as I passed the final examination and entered Khalsa College in 1955 for a two-year intermediate program, which included science, math, Hindi, and English. I was now a young man with an eye on the future. I began to focus on becoming a medical doctor. Although I truly enjoyed math, I was dismayed to learn that it was not a requirement for medical students. In 1957, I was accepted into medical school, with the idea that my future was now secure and that I was about to embark on a commitment to the medical profession.

A Change in Plans

My father had unknowingly set into motion a series of events that would change my life. By 1955, just before I entered college, my father had worked hard and managed to move our family out of the refugee camp and into a two-bedroom apartment, about a mile from my college.

It was there that my path began to change, because we were fortunate to live next door to a good neighbor (a Gujrati) who happened to have been educated in the U.S.

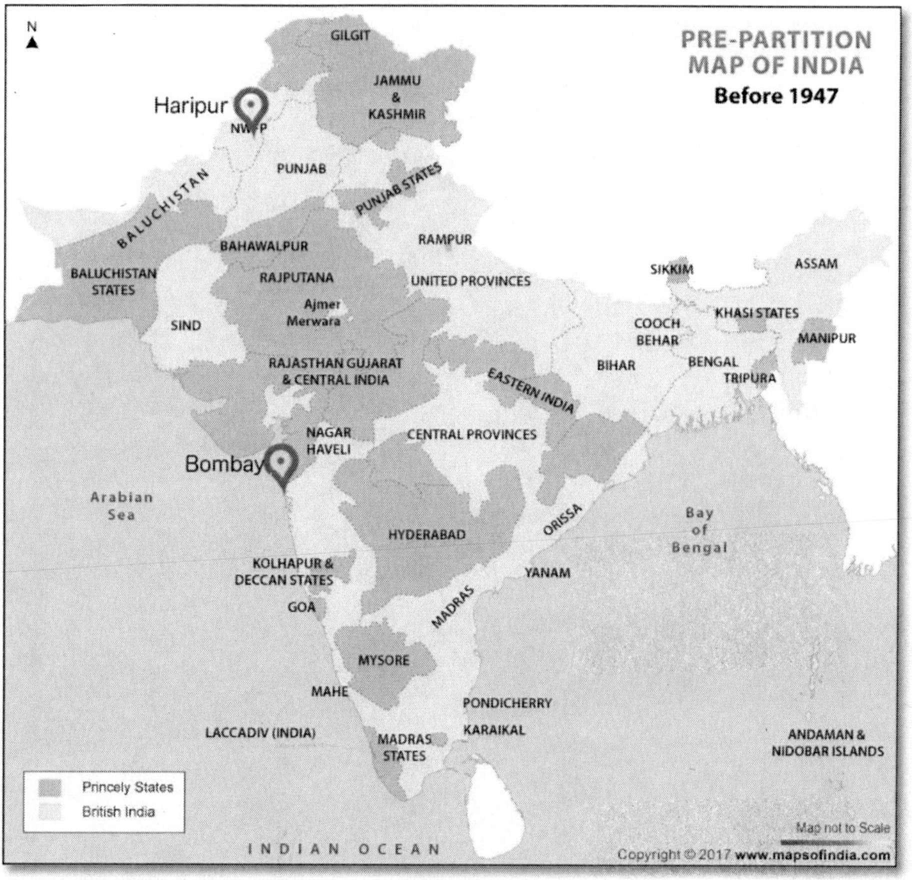

FG-1-1 Pre-Partition Map of India before 1947

My parents with my younger brothers Shree and Raj in 1957

Chapter 2
Higher Education

At the time, it appeared that my path in life was assured. My parents were determined that I should pursue medicine, so I enrolled at the medical school in Bombay. I was admitted two weeks after the classes had already started. While the idea of helping others was powerful, I did not find that medicine was particularly attractive to me. My mind kept wandering to those theorems that I used to write on the dusty streets. I began to form a new plan.

My parents had allowed my older brother, Madan, to attend school in the U.S. He was doing quite well in his doctoral studies in mathematical statistics at the University of Illinois at Urbana-Champaign. I decided that he might be able to help me in my career path. I sent him my grades from my intermediate exams; to my surprise and delight, he did more than just think about what I might be able to do. He took matters into his own hands.

After receiving my grades, Madan went straight to the people who could best help me— the admissions counselors at University of Illinois. After discussing my grades and aptitudes, an irresistible alternative to medicine was proposed to me: chemical engineering. The admissions counselors even offered me admission to the undergraduate chemical engineering program. The real challenge now was to persuade my parents that I should join my brother overseas; I knewit would be no easy task.

My professor at the medical college was ready to help me; he viewed the U.S. as an advanced nation where I could be exposed to cutting-edge research and technology. My parents, on the other hand, had a somewhat different perspective. After all, they already *had* a son pursuing a higher education degree in the U.S.; what they really wanted was for me to open my own private practice after graduating from medical school. Persuading my parents otherwise was very difficult, but I believe they were influenced by their good relationship with our neighbor and the positive experience my brother was having in the U.S. I was extremely excited when they finally gave their consent. I went ahead and got my passport and visa; but I had to face another hurdle—I had no money.

Humbly, I went to my parents and asked them if they could pay for my travel; there was no one else to help me. What a thing to ask! My loving parents understood me well and agreed to pay for my trip. My future was now beginning to gather momentum. I sent a short note to my brother that I would soon be joining him. Who knew what I would find in the U.S.? The possibilities seemed endless. I could not wait to see my brother and start a new life.

Chicago

I flew to Chicago the first week of September 1957. In the 1950s, the U.S. was in a time of change. Civil rights struggles were in the headlines; President Eisenhower was already deep into his second term of office and had signed a new civil rights bill; Buddy Holly and the Crickets were a sensation; Borazon, a new material that was the second hardest material after diamond, was created in a lab; Basil Hirschowitz began transmitting light via fibers, leading to fiber optics; and Irving Langmuir, the 1932 Nobel Prize winner in chemistry, had passed away, shortly before I first set foot on American soil.

The heat of the Chicago summer had abated, and the weather was relatively pleasant. Chicago was a long way from Bombay; I was in a completely different land with many new opportunities. I took a train to Urbana and watched as the miles between me and my future were devoured. A unique world passed by outside my window; a world that was being recreated even as I entered it. I was not aware that, in January, the university had hosted a U.S. senator from Massachusetts who had given a stirring speech about politics and that I had missed the initial outbreak of the infamous "water fights." But I would get a chance to hear from that senator again, when he, John F. Kennedy, was sworn into the Oval Office the year I completed my undergraduate studies. Like the new country I now called home, my life was changing rapidly, and my heart was full of hope.

Life in Urbana

I arrived in Urbana on a Sunday and spent a bewildering time at the station. It was not that I was lost, but simply that I was not found. I desperately needed to find Madan, but he was nowhere to be found. I didn't know what to think. I made my way to the main campus. I had only four dollars with me, at that time the maximum allowed by the Indian government for foreign travel. Even in 1957, this was not a lot of money. A gallon of milk cost about $1, a loaf of bread was 19 cents, and the rent for a small apartment averaged $186 a month.

I somehow managed to find my brother's apartment, which was located off campus. By now, I was weary and hungry and totally unprepared for the next surprise—the apartment was locked and empty. What had happened? I kept walking around. Where was Madan? I *had* to find my brother, no matter where that led me. The best place to look for any Ph.D. student in the 1950s is still the very best place to look today—on campus. I managed to navigate my way to Madan's office, which was occupied by that gentle face that I had known so well! Madan was in shock when he saw me at his doorway. After a joyful reunion, I understood why my brother had not come to the station—my letter had never reached him.

By now, I had to ask Madan a question that had been burning inside me. I had ended up at the University of Illinois and enrolled in chemical engineering, though I had no idea what chemical engineering all about was. Why had my brother chosen it? What made him decide on that path for me? Quite simply he had asked an expert! He had taken my transcript to an admissions counselor, who suggested that I choose this particular major. I let this novel piece of information roll around in my mind, finding it striking that people we didn't know could have such enormous influence on our lives. An anonymous counselor, who had reviewed an equally anonymous, poor, young man's grades, had instantly opened a door to endless opportunities for him!

Madan's apartment was located near the campus—a tiny studio apartment that he was sharing with a roommate from India.

The roommate was an interesting person; a Punjabi, he had obtained a coveted Government of India scholarship to study agricultural economics. Our sleeping arrangements were simple; my brother indicated that I was lucky to get the top bunk. He and his roommate took turns cooking our meals. I didn't know how to cook, which meant that I was often in charge of washing the dishes.

Taking Courses

There was another apartment on the same floor occupied by a doctoral student studying in my brother's department. A friendly person, he had formerly been a math professor in the state of Kerala in India. He took it upon himself to mentor me in calculus—an incredible kindness, since I had not taken *any* math courses in my second year at college in Bombay. I was not well prepared to take calculus, and I was wholly unprepared for analytical geometry. I was very nervous the entire first semester. Moreover, I was homesick. It had only been about a month since I had moved to the U.S., and I was on my own, with no friends, no ecosystem, and little to no outside interests. Just my brother and my work, and I truly missed my mother's cooking. I had no idea what "American" food was, because I never ate out. Clearly, budget was one factor, but we were comfortable making our own food, and it was predominantly Indian. However, I would take 10 cents every week and run to the Illini Student Union, so I could get that coveted ice cream cone that I so loved. Yes, it was simple living, but it worked for me and taught me many lessons on how to appreciate the little things in life.

I had registered for two five-credit-hour courses, calculus and quantitative analysis in chemistry, and I was required to take a six-hour-per-week, noncredit course in English rhetoric. The U.S.

education system was different from what I had grown up with, a system based on memorization, where only the final exam grades were considered. The new system required critical thinking, doing homework regularly, and taking lots of quizzes. I studied very hard during my first semester, and, luckily, I received as in all my classes. The chemistry class naturally involved labs (four hours a week), but in reality, I actually spent at least twelve hours a week. I enjoyed the lab work and my teaching assistant, who allowed me to spend my free time puttering around in the lab.

Pleased that I had received all As, my brother strongly urged me to apply for a tuition scholarship. He told me to go see Mr. Schneider, the dean of foreign students, who soon relayed disappointing news: because the University of Illinois was a state school, no tuition scholarships could be awarded to non-native undergraduates. However, he promised me that he would check with my teachers. Shortly before the beginning of the second semester, and incredibly, I was informed by his office that I was awarded a full-tuition scholarship for the next two-and-a-half years, provided that I maintained a B+ average.

I was overjoyed, and joy is always multiplied when it is shared; Madan was perhaps even more excited than I was. At that time, tuition was $250 per semester and $7 for health insurance—no small sum for a poor doctoral student. Although my brother had paid for everything for my first semester, the best news was that I could now graduate after the remaining two-and-a-half years. I had to attend summer school, but the tuition scholarship covered the summer session as well. Moreover, the university also gave me transfer credits for the science courses I had taken in Bombay.

Chemical Engineering

Founded in 1901, the University of Illinois's chemical engineering department is one of the oldest and most established programs in the U.S. I soon found that it was brimming with excitement and fresh ideas, and I had the exceedingly good fortune to study under greats like Tom Hanratty, John Quinn, Jim Westwater, and Harold Frazer Johnstone. At the time, Max Peters, a decorated World War II veteran who would later go on to serve as dean of engineering at the University of Colorado, Boulder, was the department chairman. Harold Johnstone, the former chairman, was a research professor. Johnstone was not engaged in any teaching, but still ran an active research program in fine particle technology related to air pollution.

Professor Hanratty had a passion for his work. His contributions to the understanding of fluid mechanics can be seen in many applications today, such as the design of the Alaska pipeline. Defying the stereotype of the introverted scientist, he strongly encouraged his students to not only present their findings at professional meetings and conferences, but also to highlight the importance of interacting with scholars from various other research facilities. In 1958, during my first summer, Hanratty hired me as a full-time lab assistant to work with one of his doctoral students, Hisao Kada, who was conducting research on turbulent diffusion. I was so intrigued with this topic that I later chose it to be my doctoral dissertation at University of California, Berkeley.

As summer tumbled into fall, Kada, a very kind man, recommended that I stay on at the lab. Some things never change, though, and this is especially true for lab assistants. I was paid the princely sum of $1 per hour for my lab work, while the average wage for a factory worker was $1.75 per hour. In addition to my full-time lab work, I took two courses during each of the three summers so that I could graduate on time. During that summer, my brother and I moved from Urbana to Champaign into a new apartment. Coincidentally, 30 years later my son Ajay lived in the same apartment when he attended the University of Illinois! He is now vice chair for pain medicine and a professor of anesthesiology and psychiatry at the medical college of the University of Pittsburgh.

Things were going well. My grades were up, I had an absorbing job, and the people around me were wonderful. However, there was a bump in the road—my class in material and energy balances (stoichiometry). It was my first course in chemical engineering, and I found it disappointingly difficult. I worked very hard to maintain my scholarship, but I earned only a passing grade. The fall semester of 1958 was a bleak time in my life; however, it only made me buckle down and study even more. I performed better in my subsequent courses, and I obtained excellent grades in math, my academic joy. My studies were progressing so well that I was awarded a $200 scholarship from the department.

During the summer of 1959, opportunity came knocking; although it was paired with some sweet sorrow. Kada had completed his studies and was on the verge of graduating—thus my assistance was no longer required. The thought of being out of the lab was not pleasant at all, but I had to face the facts. Fortunately, however, this was at the time when Hanratty introduced me to Professor Johnstone; I think Hanratty felt that I was quite able to do some independent research under Johnstone.

Working with Professor Johnstone was a revelation, an experience that had a powerful impact on how I would approach research in the future. I worked full-time in his lab over the summers of 1959 and 1960 and part-time during the 1959–60 academic year. I was assigned to a project on aerosols, with their applications to acid rain. I used to meet with Johnstone on Saturday mornings to discuss my research findings. I learned a lot working with him, as he was a very accomplished engineer and scientist. He taught me how to conduct duplicative experiments and analyze data, using statistical methods, before drawing any conclusions. Luckily for me, I already had taken a course in statistics and was familiar with statistical analysis.

At the end of my second summer, my brother wrapped up the requirements for his doctoral degree in statistics. He was offered a faculty position at Queens University in Canada. Although I was sad to see him go, I was very happy for him. Madan, for his part, had always looked after me, even when he was getting ready to move to another and more exciting stage in his life. Just before leaving for Canada, he made arrangements for me to live with another roommate. I moved from Champaign back to Urbana and shared a new apartment very close to the lab where I worked.

My days of working in the lab at the University of Illinois were soon coming to a close, and I made a decision to enter graduate school. I was, however, somewhat concerned about my future. I had a good academic record, maintaining a B+ average, but I wasn't sure if it was good enough for admission to the University of California, Berkeley, whose reputation for excellence was renowned. I asked Professors Johnstone and Hanratty for recommendations. Although I was not sure of what would happen, I applied, hoping for the best but prepared for the worst.

Waiting for Berkeley's decision was a period of great uncertainty; like many students who hoped to be accepted, I was unsure of what exactly would happen. When I received my letter of acceptance, I was filled with joy and relief, and I am very grateful to this day, for I know well that the recommendations from Professors Johnstone and Hanratty opened a door for me.

Berkeley

In August 1960, I began to pack my few belongings for my journey across America to the Golden State. I knew that life at Berkeley would be a change from what I had known at the University of Illinois. I found a roommate, from India, and we shared an apartment near the campus. He was gracious enough to share the gifts of his time and talent—he taught me how to cook!

I served as a teaching assistant in unit operations, a lab course for seniors. Professor John Prausnitz, known for his research in molecular thermodynamics for fluid-phase equilibria, was the instructor. He showed great faith in me, and I did my very best. I initially graded a few lab reports submitted by the students and showed them to Prausnitz. He was so pleased with my evaluation of the reports that he asked me to conduct the labs independently, without the need for checking with him. This simple expression of faith opened my eyes to a new possibility: What if I could become an instructor myself? It was a pleasing, encouraging thought, the idea that I could be helping others achieve their hopes and dreams!

However, my path through academia was going to be more challenging than I initially thought. During the fall semester, I took the Ph.D. qualifying exam. I passed, but with one caveat: I had performed poorly on the thermodynamics section of the oral exam. Fortunately, I was asked to serve as the teaching assistant for the undergraduate thermodynamics course that Professor Andreas "Andy" Acrivos, now well known for his work in fluid dynamics and transport processes, was teaching in the spring semester. He was a great teacher, and it was this course that helped me truly understand some of the most basic thermodynamic principles. Just before the summer of 1961 began, I managed to pass the second qualifying exam. This exam was a bit of a challenge, because I was required to defend an original research proposal of mine—an experimental method to investigate aerosol coagulation kinetics. My success on this exam was due, in no small part, to the training in colloid science that I had received under Johnstone. I also passed exams in two foreign languages (requirements at the time), French and Russian, both of which I had taken at the University of Illinois.

During my first year at Berkeley, Professor Johnstone came to see me. He happened to be visiting his former doctoral student, Tom Baron, and was interested in seeing my progress. Tom was the director of the research lab at Shell Development Corp. I was, of course, very pleased to see Professor Johnstone, and he was glad to learn that I was about to start on my doctoral research and that I was going to work with Professor Charles Wilke as my research advisor.

Although I had studied under Andy Acrivos, John Prausnitz, Donald Hanson, Charles Wilke, Donald Olander, and Chang-Lin Tien (who later served as chancellor at Berkeley), I chose to conduct my doctoral thesis research work under Professor Wilke, one of the founders of the chemical engineering program at Berkeley and a pioneer in biochemical engineering in the field of mass and momentum transfer in turbulent flow. I was assigned a lab for my research work at the Lawrence Radiation Laboratory (Lawrence Berkeley Lab, LBL), located on the hills near the campus.

It was around this time that I began publishing papers. I co-authored two papers on turbulent transport ("Law of the Wall") with Tien, who was teaching mechanical engineering. Life at LBL was very busy; I carried out both theoretical and experimental studies and even built a wind tunnel to carry out the experimental work. A very able machinist, G. G. Young, worked in the machine shop next door to my lab and helped me build the wind tunnel. I also numerically solved the differential transport equation, using the newly minted FORTRAN language on an IBM 704 computer. It was, at the time, a very advanced system utilizing vacuum tubes; it could process 4,000 operations per second, and the data could be put in using both punch cards and magnetic tapes.

The LBL was a federal lab, with dedicated researchers and amazing facilities. There were several very bright doctoral students working for various professors in my lab; some of them were graduates of Princeton, Caltech, and MIT. There were also two master's students of Wilke's working in my lab, Fred Heinzelman and Rob Davis, with whom I co-authored my third and fourth papers. I greatly enjoyed my time at LBL; those were happy years (1961–4) and incredibly productive as I managed to publish four papers before completing my experiments.

It was at LBL that I created my new name. One day, Professor Wilke mentioned that the name "Darshanlal" was too long and difficult to pronounce. He had understood that my first name had significance, but he asked me if I could shorten it without losing its special meaning. I was happy to comply with his request. Many people are not satisfied in truncating their name, because they feel that the essence of its meaning is lost. I, however, lost no time in shortening mine to Darsh; after all, in Hindi, it has exactly the same meaning as Darshanlal. Both names relay an idea that has been central to my life—vision!

A Different Campus

Looking back at the 1960s, I had not sensed that the Berkeley campus would shortly erupt into riots and become part of the national discussion on free speech and the counterculture movement. During the four years I spent there, it was a peaceful campus. In fact, I only learned about the riots and other related events through letters from some of my former, fellow Berkeley students after I had moved to Chicago. When I visited Berkeley again, two years after I began my career, I was stunned to see the abrupt transformation. There were changes to the campus and its surrounding neighborhoods; the students and hippies were now sitting on the sidewalks of the city streets.

Job Interviews

In the spring of 1964, I began interviewing with a few of the companies that had come to recruit students. Three companies on the East Coast, Esso Production (now Exxon), Union Carbide, and American Cyanamid, asked me to come and visit them. I went to see Professor Wilke to inform him about these visits; he told me to stop over in Chicago and visit Professor Ralph Peck, head of the chemical engineering department the Illinois Institute of Technology (IIT).

Apparently, Professor Peck had asked Wilke to recommend someone who had worked with him on mass transfer. I was not quite sure what to expect, so I wrote to Peck; he, in turn, invited me to present a departmental seminar as a part of my job interview at IIT. I rehearsed my lecture with my fellow graduate students; because I had an accent, they advised me to speak slowly and clearly. I invested in a small tape recorder and recorded my speech in order to improve my delivery. The input that I received from my lab mates proved to be invaluable, and I ended up delivering a good presentation.

IIT offered me an assistant professorship in the College of Engineering, beginning in the fall of 1964, at an annual salary of $8,000 (about $70,000 today). Still, I continued on to the East Coast to visit the three companies. After lengthy interviews, I was delighted to receive offers from each company, all at an annual salary of approximately $12,000 (about $105,000 today).

A career is always about more than just money; the personal satisfaction you derive from your work and the people you can influence are crucial to your choices. I now stood before many open doors, but I realized that I needed additional input. When I returned to Berkeley, I discussed my offers with Professor Wilke, Vice Chairman Donald Hanson, and several of my fellow graduate students. All of my peers recommended that I pursue an academic career; they felt that I had delivered well-prepared and enthusiastic lectures. Hanson also sided with my peers, because he had received positive reviews on my assignments as a teaching assistant, and he thought I could be an effective instructor. While the money was certainly enticing, I ultimately decided to follow my heart.

Although I had one obstacle out of the way, choosing to work at IIT was the easy part. I still had to complete my experimental work at Berkeley, write up my dissertation, and defend it, all before moving to Chicago in early September. With my head full of my future responsibilities at IIT, I began to race along that rocky road well known to all doctoral students. The summer of 1964 found me sweating in the lab, from 7 p.m. to 7 a.m., seven days a week. I took the final oral defense exam and left my final original copy of my thesis with Professor Wilke to file in the graduate school so that I could graduate. In one hectic summer, my life as a student had come to an abrupt end.

I was now entering the world of academia as one who was ready to help and guide others; it was a humbling thought. But there was much yet to do for my move to Chicago. I contacted a travel agency to book my flight. When I went to pick up the plane ticket, I received a surprise— Professor Wilke had already paid for my ticket! I was overcome by his generosity. When I went to thank him, he kindly offered to drive me to the Oakland airport.

I will never forget that morning. Always true to his word, Wilke picked me up and drove me to the airport. As I was about to say goodbye, he did something much unexpected; he removed his coat and placed it in my arms! “Winters in Chicago can be very harsh,” he said, “You will definitely need something warm.” I was deeply moved by his kindness and thanked him profusely. Wilke was indeed right; Chicago winters could be very severe. That overcoat saw a lot of use for the next several years. Professor Wilke and I remained in touch for the rest ofhis life.

The Farewell

Professor Wilke’s acts of kindness remain etched on my mind, often stirring up the many good memories of my interactions with him. His mentorship, graciousness, and care for his students made an indelible mark on me. His passing in 2003 at the age of 87 was a great loss, not only to the scientific community, but especially to me and to the legion of lives he had touched. He had always been a father figure to me.

In the fields in Urbana in 1958

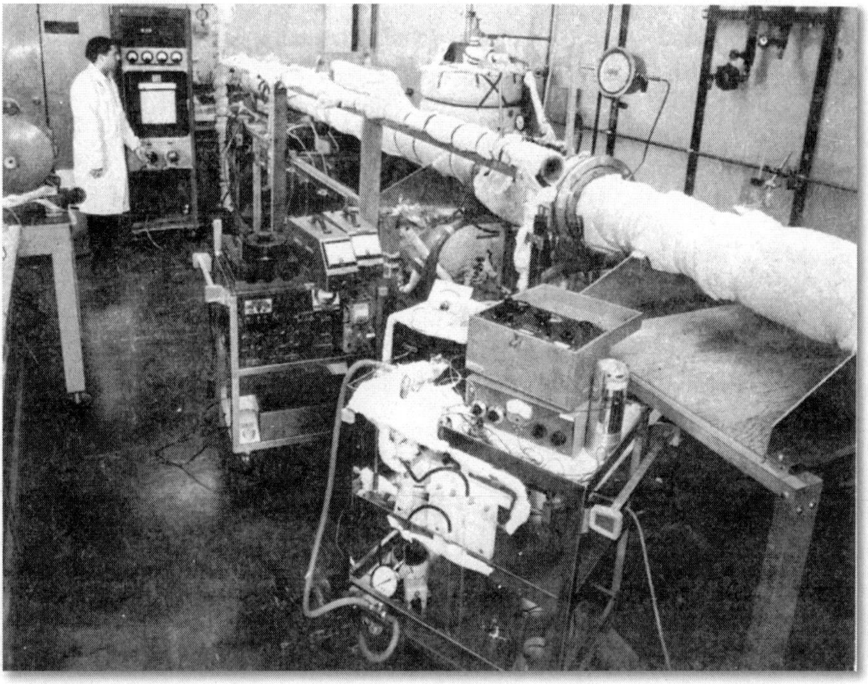

In my lab at Berkeley: I build this experimental set-up to complete my doctoral research.

Chapter 3
IIT: The Early Years (1964–71)

In academia, the fall semester is the beginning of a novel adventure in learning. Incoming freshmen are just beginning to explore the freedom and promise of university life, and graduate students are recalibrating themselves for a deeper, richer understanding of their fields. Instructors who have been hard at work over the summer with field research, grant writing, supervising students' thesis defenses, and prepping for the upcoming semester are energized by the thought of educating their students and fresh challenges. This generates a special kind of excitement that is bolstered by the changing of the seasons and cooling autumn weather. I was about to become a part of this excitement.

Year One

I arrived in Chicago on a Sunday in early September 1964, just as the leaves were beginning to turn gold and cooler breezes began to waft in from the lake. Like many single young men, I arrived with little, but I was not concerned about owning things; my passion lay entirely in my work. I was eager to enter into the unexplored realms of engineering science, enticed by the challenge of nature.

I moved into a studio apartment at Carman Hall, which had been housing IIT's students since 1953. Carman Hall was one of about twenty buildings on the IIT campus designed by Ludwig Mies van der Rohe, a German architect who served as the head of the architecture school from 1938 to 1958. (S. R. Crown Hall is often considered the epitome of Mies' architectural career.) Mies was internationally acclaimed for his dedication to modernist design. While he was quite well known for the Chicago Federal Center Complex (now the John C. Kluczynski Federal Building) and the National Gallery in Berlin, his most important work was education. Mies was dedicated to the aspiring architects in his care and often worked with them on projects.

Mies' architectural footprint would have surely bypassed Chicago if it had not been for a preacher and a millionaire. IIT had originally been the inspiration of a simple Chicago minister, Frank Wakeley Gunsaulus; he is probably best remembered for delivering what became known as the "Million Dollar Sermon." During his impassioned preaching, he promised that he could create an educational institution that would serve the needs of regular people—if only he had $1 million. Philip Danforth Armour, a multimillionaire, took Gunsaulus up on his bargain and invested that exact amount to create the Armour Institute. In 1893, the Armour Institute opened its doors to students and offered classes in engineering, chemistry, architecture, and library science. IIT fully materialized in 1940 when the Armour Institute merged with the

Lewis Institute, which offered a range of courses in the liberal arts and sciences. Later, under the leadership of President Henry T. Heald, IIT became an essential part of the war effort when it began offering military and civilian training courses in the 1940s. This expansion was so significant that IIT became known as the war training center of the Midwest. The end of the war did not halt the continuing expansion of the school, though. The Institute of Design merged with IIT in 1949. By the time I arrived in 1964, IIT had become a center of innovation and learning.

So on a Monday morning just before the fall semester began, I went to the Department of Chemical Engineering in Perlstein Hall and met with Ralph Peck. Peck introduced me to the academic I was hired to replace, Professor Bob Kintner. Kintner, who had been teaching and conducting research on mass transfer, was going to retire the following year. At that time, the chemical engineering department encompassed a small but dedicated band of researchers. There were only five tenure track faculty members: Professors Peck, Kintner, Swanson, Levenspiel, and Weinstein. The department also had two nontenured instructors, John Marek and Steve Szepe.

By the 1960s, IIT was well known for its engineering education and research. One of its electrical engineering graduates, Marvin Camras, who invented the magnetic tape that underlies most of the electronic and digital media, was still actively working at the IIT Research Institute (IITRI). At that time, the IITRI had over 1,000 employees at 16 national sites. While the IITRI was founded in 1936 as the Armour Research Foundation, the tower that most IIT students recognize as the IITRI only went up in 1965. Now as then, the IITRI has always had a hand in the most cutting-edge research involving both private companies and the government.

John T. "Jack" Rettaliata, an eminent mechanical engineer who was well known in the field of steam and gas turbines, was the president of IIT then. The Institute of Gas Technology (IGT), founded in 1941, was also on the IIT campus and was closely affiliated with the university. Its president at the time was Henry R. Linden, who was a former doctoral student of Peck's. (Linden also served as the interim president of IIT from 1989 to 1990 after President Meyer Feldberg left to become the dean of New York's Columbia Business School.) The IGT employed approximately 200 people, and some of the IGT employees were part-time students at IIT, primarily pursuing their graduate degrees in chemical and mechanical engineering. Several of the graduate students at IIT were financing their studies by working part-time at the IGT.

During my first week at IIT, Ravi Randhava, a new Ph.D. student from India, knocked on my door. I had met his father the past summer at Berkeley and had mentioned to him that I was joining IIT as a faculty member in the fall. Randhava informed me that he would like to join my research group. I was eager to work with Randhava, but that age-old problem crept up again— money. Randhava needed some financial support to begin working, but I sadly informed him that the department had not allocated any startup funds for my research. Life, however, is full of strange twists and turns. One week before classes began, we had our first departmental faculty meeting. It was at this meeting that Peck mentioned there was a single teaching assistantship open and asked if we could suggest some names for this position. It was exactly the opportunity I was looking for, and I immediately suggested Randhava, who was awarded the teaching assistantship shortly afterward.

I was preoccupied now with my first semester courses. I was assigned to teach three courses, including a unit operations lab for juniors that was scheduled on Saturday mornings. I felt especially prepared for this class, since I had taught a similar lab at Berkeley and had enjoyed doing it. My two other courses included a unit operations lecture class for juniors and a graduate course on the principles of diffusional unit operations. I kept myself very busy preparing detailed notes for the classes that I taught. I wanted to not only identify the best ways to teach my students, but also discover the types of problems they encountered with understanding the material.

While my classes took up most of my time, I did not want to neglect my passion for engineering research. I managed to prepare a research proposal, entitled "High Flux Mass Transfer in Turbulent Pipe Flow," that I submitted for possible funding to the National Science Foundation (NSF). Randhava offered his invaluable assistance to me in the preparation of this proposal, which we sent off with some trepidation to the NSF. And, just as the position had suddenly opened up for Randhava, the NSF came through for me. My proposal was funded just before the start of the 1965 summer semester, and I received a research initiation grant of $15,000 (over $100,000 today) for two years. This grant funded one month of my summer salary, stipends for my research students, and even provided money for some supplies for our experimental work. The funding also allowed me to support Randhava during the summer and for the next year as a research assistant.

My tiny research team began to expand. Another Ph.D. student, Bill Jones, and a master's student, Garry Von Behren, would be the first students to join my team that summer. Bernie Baker, who had taken my graduate course on diffusion and was managing the electrochemical engineering group at the IGT, decided to conduct his doctoral research on heat and mass transfer in electrochemical systems under Professor Dimitri Gidaspow's and my joint supervision. At that time, Gidaspow was teaching in the IGT's gas engineering education department. He had earned his Ph.D. in chemical engineering at IIT a couple of years before I arrived. Little did I know at the time that Dimitri and his amazing wife, Helen, would become dear friends to my wife, Usha, and me and our family for over 50 years.

While life had been humming along for me, I hit a bump in the road during my spring semester in 1965. It was then that I was asked to teach an additional graduate course on non- Newtonian fluids. As a graduate student at Berkeley, I had learned about classical fluid mechanics, but I had very little knowledge of non-Newtonian fluids. While I certainly understood the basic concept behind these types of fluids (as does any child who has made Oobleck in their kitchen, much to their parents' vexation), most nights found me diligently learning new material and deciding how to best teach this course.

My hard work paid off, because it resulted in capturing the attention of another curious mind—Anton J. "Tony" Pintar. Pintar, who would later go on to teach chemical engineering at Michigan Tech for 36 years, decided to join my research group after taking this class. He had a wonderful love and aptitude for mathematics, and was already receiving a fellowship from IIT, so he did not need any financial support from me.

There was much to do my first year at IIT. Fortunately, I could keep long hours in the office since, at that time, I lived on the campus and was still single. However, in 1965, my younger brother Shree came to live with me. I was excited to have him stay with me in Chicago, but I was quite busy with my work at the university. Eventually, I was able to help him get admitted to a junior college located on the South Side of Chicago. Shree lived with me for nearly two years before deciding that he needed to expand his horizons. He moved to the East Coast and ultimately earned his bachelor's degree in electrical engineering from a college in Connecticut. (Shree is now retired and lives in Dallas, Texas.) I am afraid that my younger brother may have felt somewhat neglected at times. I vividly remember that there were days when I would be in my office until 3 a.m. advising my research students and then return to teach my classes at 9 a.m. This, however, was never a problem for me; I was doing what I loved and making a real difference in the lives of my students.

IIT's Investment in Tomorrow

In the 1960s, IIT was considered one of the top three private technological universities, along with MIT and CalTech. Our school was especially known for preparing young people for the types of work they would go on to do after graduation; the school had always placed a lot of emphasis on hands-on laboratory work. Because of the strong focus on practical learning, IIT had considerable interactions with industry in the Chicago area. We had courses full of dedicated, professional part-time students from different industries attending the evening classes. Most of these part-time students paid full tuition, so this proved to be a good source of tuition revenue for the university.

However, supporting students and exploring the vast areas of the unknown always include considerable expenses. Funding is a thorny issue. IIT was founded with the goal of helping those who were not born into elite families attain the skills they needed to achieve their professional goals. President Rettaliata took this to heart, and in the spring of 1965, he embarked on a fund-raising campaign like no other the university had ever seen. Twenty-five million dollars were committed to his "Investment in Tomorrow" campaign, and he even secured a coveted $5 million challenge grant from the Ford Foundation.

This is how IIT underwent a name change. After the Armour Institute merged with the Lewis Institute, the new university was called "Illinois Tech." It was only in 1965 during President Rettaliata's Investment in Tomorrow campaign that we became known as "IIT"—to better reflect the fact that we were a research leader like MIT. In recent years, the university has returned to its roots as Illinois Tech, but the commitment has always been the same. A rose by any other name does indeed smell as sweet.

During the summer of 1965, IIT hosted the first World Congress on Engineering Education, which was held in the newly built Hermann Hall. The dean of our engineering college at the time, Ralph G. Owens, and the chair of our mechanics department, Peter Chiarulli, were quite involved in organizing this conference and asked me to serve as a volunteer. I, therefore, had the pleasure of meeting some prominent engineering educators from around the globe. About 600 attendees were present at this international gathering.

The IITRI building was finished and in use by 1965 and was formally dedicated in 1966. Designed by Schmidt, Garden and Erikson, the 20-story tower was the tallest research building in the Midwest. Research at the facility included both classified and unclassified projects in diverse areas such as physics, computer science, astrophysics, and electronics. The dedication of the building took place at the end of April in 1966. It was a monumental event. Both Illinois Governor Otto Kerner Jr. and Chicago Mayor Richard J. Daley participated and helped cut the tape (which was magnetic recording tape, of course!) for the new $10 million facility.

Year Two

In the fall of 1965, Chang W. Chi, another Ph.D. student, joined my research team. Chi had taken my graduate course in diffusion the previous year and was piqued by the opportunity to advance his studies. He had decided to work under my supervision for his doctoral research, which he carried out in the evenings and weekends at the IGT while being employed there full- time. Beginning in the fall semester, several new master's students, including Dharamvir Punwani, Narayanan Sankar, and Narayanaswami Srinivasan, were added to the group. Teddy Schmidt, an IGT employee, and Kushalkumar Baid also joined my research group. I advised Punwani to work closely with Chi on his master's thesis project at the IGT, which was providing funding for him.

During my second year, I began to meet new people and expand my horizons. Joe Kaellis, an Argonne National Laboratory employee who was working on his Ph.D., joined my research group after taking my graduate diffusion course, and two new Ph.D. students joined my research group as well. I also started interacting with the scientists and engineers in the fine particle section at the IITRI. Al Lieberman, who was an expert in aerosol science, was the section manager, and Brian Kaye, a physicist and prominent particle technologist, worked in this section. They were trying to grow the fine particle team and asked me to work closely with them. I thought this was a great opportunity for me to pursue my interest in colloids that had been stoked by Johnstone when I had worked on an aerosol project under his supervision at the University of Illinois.

Reg Davies was recruited in 1966 by the fine particle section at the IITRI. Davies was a chemist but had worked in particle technology in England before joining our research institute. Davies and I started interacting from the day he arrived on our campus. He was planning to earn his Ph.D. by conducting his research work at the institute and submit the research thesis to Loughborough University in England, which would grant him the degree. Due to a university requirement, Davies needed an external thesis advisor, and I was very pleased to serve as his advisor here and started working closely with him on his research work on fine particle processing. Also, that year, Davies, Kaye, and Lieberman started preparing a research proposal on particle technology to seek funding from approximately 15 different industrial organizations. I accompanied Davies on many of his visits to these companies to discuss problems of interest to them. Subsequently, the IITRI hired me as a consultant and I was paid $200 as a one-day consulting fee; this consulting opportunity eventually evolved into two days a month. At about that same time, the IGT hired me at the same rate to consult for their electrochemical engineering group one day a month. The IGT also funded Teddy Schmidt's master's thesis research work on mass transfer in fuel cells under my direct supervision.

Davies made a great name for himself in the field of particle technology while working at the IITRI; he received his Ph.D. from Loughborough and eventually was recruited to work for DuPont. In 1973, Davies and I conducted a national workshop in Philadelphia on the needs and priorities in particle technology. This was attended by many academics and industrial scientists and was sponsored by the NSF, which published its proceedings in 1975 as a national document (NTS No. PB267837-AS). The NSF had started a new program section that would fund research on particulate and multiphase processing. The program director had mentioned to me on several occasions that our proceedings document outlining the research needs and priorities in this subject area served as a primary reference source for him to fund proposals for years to come. Davies and I later took turns serving as the president of the Fine Particle Society before he joined DuPont. Davies also taught a new course on particle technology in our department when I became chairman in 1971.

By the end of my second year at IIT, I was mentoring eight doctoral and six master's students and already had one NSF grant to support some of them. I was thinking ahead about venturing into new research areas of interfacial and colloidal phenomena in chemical engineering. I felt these new areas had a broader industrial appeal and would build upon the colloid science background that I had gained studying under Johnstone. In contrast to the field of turbulence, this subject area did not require using big expensive equipment, such as wind tunnels, and it was also an area that was in the mainstream of chemical engineering. I first proposed this idea to Pintar for his doctoral dissertation. I envisioned the project would entail building a simple shear viscometer to measure the surface shear viscosity at the gas/liquid surfaces containing surfactants or macromolecules. The first part of the project required developing a mathematical analysis for designing and building such an instrument and then conducting experiments and analyzing the measurements. Pintar was an excellent researcher and had a strong background in applied mathematics. Within a year, he was successful in completing the theoretical analysis for the design of the surface shear viscosity, a 2D analog of the 3D bulk viscosity. The next step for Pintar was to build this apparatus based on his theoretical analysis and to analyze the data. Luckily for him, a postdoctoral Rotary Fellow from Spain who was an experimental physicist had joined my research group that summer and assisted Pintar in building this instrument. Pintar collected data on a few systems using this new device and completed his research work by the end of the summer of 1966. With Pintar's experience and skills, he was soon scooped up by another school; in the fall of 1966, he left IIT to teach at his alma mater, Michigan Technological University.

A Visit to Bombay

During the summer of 1966, I decided to go back to Bombay to visit my parents, whom I had not seen for nine years. I hardly got to speak to them because you had to book the call eight hours in advance. It was a different time back then; without the Internet and cell phones, overseas communication was mostly done by writing letters. However, when I passed my Ph.D. exam in 1964, I did call my parents, and they were definitely elated.

But I did not pack my bags for my trip home right away. Research team members become a very closely-knit unit. We do, after all, work very hard to achieve the same goal in a limited amount of time; we must rely on each other, just like the members of a family. I put off my own trip home so I could attend Chang Chi's wedding.

I left for India at end of June, but I took a circuitous route: I went to India by way of Japan. This was because I had been invited to Kyoto University to present a graduate seminar in the chemical engineering department. I had known Professor Fumitake Yoshida at Kyoto from my graduate days at Berkeley, which was when he had asked me to visit him. At Berkeley, he served as a visiting professor, and I assisted him in teaching his graduate course in mass transfer.

I finally arrived in Bombay in July after almost a decade of living in the U.S. My parents were overjoyed to see me—and introduced me to a beautiful Punjabi girl, Usha Kapur, who held a bachelor's degree in history. Actually, I was introduced to five girls through my parents. That was the process of finding a wife back then. I did not have the convenience of Internet dating. I chose Usha (or rather we chose each other) because she was educated in arts, which very nicely complemented my scientific background, and also, she had always dreamt of making a home abroad. Her father was an engineer. I met Usha's elder brother when my family first went to meet Usha at her house on August 3, 1966 and found that he had studied at the school in the refugee camp—Guru Nanak—while I had been a student there.

Newlyweds

Usha and I were married on August 21, 1966, and then we were off to an interesting honeymoon. We first flew to Brussels, where I presented a paper about non-Newtonian fluids at an international conference on industrial chemistry, then we spent the weekend in Paris and finally flew to England, where I delivered a departmental seminar at the Imperial College London. Usha and I flew back to Chicago from London on September 21, 1966. Several of my research students met us at O'Hare airport to greet my new bride and drive us to the campus.

I would like to say that my married life started perfectly, but there was one snag that I did not foresee—housing. Usha and I moved into a two-bedroom unit at Carman Hall, which was completely unfurnished! I had failed to request a furnished apartment before I left on my whirlwind of summer travels, and so I presented my beautiful bride with an apartment full of nothing. However, all was not lost. Randhava heard of my plight and managed to wrangle a mattress for us from his student dormitory.

Usha and I began our married life together by sleeping on a mattress on the floor for nearly two months in a faculty apartment. We ended up living on campus for the next nine years. (Fortunately, I did manage to eventually obtain other pieces of furniture!) Those early years of my marriage were so full of joy. My student Chang Chi and his wife, Choya, were staying in the apartment next door to us in Carman Hall.

Usha was a little overwhelmed with life when she came to the States. Everything was new and different for her, but she adjusted well. She joined the Faculty Wives' Club, where she soon became president. In addition, she organized book and bake sales with some good home cooking for the students, which helped raise funds for the new playground for the children on the campus.

Teaching Award and a Promotion

My third year at IIT (1966–67) was certainly one of my most memorable ones. I was honored to be nominated by both undergraduate and graduate students for the university-wide Excellence in Teaching award. President Rettaliata presented this award to me at the IIT commencement held at the Chicago Symphony's Orchestra Hall in June of 1967; the award included a cash prize of $1,000. I also received a promotion to the rank of associate professor with my tenure beginning in the fall of 1967. I truly believe that Usha brought me lots of luck.

I had been quite surprised to receive the teaching award. I really enjoyed my work, but had never realized how I had impacted others. Many years later, I was very pleased to learn that one of my former esteemed colleagues and a member of the U.S. National Academy of Engineering, Octave Levenspiel, commented in an article published in the *Chemical Engineering Education Journal* in 1992, "Darsh is an exciting teacher because he is concerned with young people, his students, and he is in love with his subject. Research ideas bubble from him, and students flock to work with him. He works them hard with evenings and weekend conferences, but they love it. I would say that close to half the students in the department did their theses with him." Levenspiel inspired the greatest respect from all who knew him. He was probably most famous for his groundbreaking textbook, *Chemical Reaction Engineering*, published in 1961 and still in use today. He taught at

IIT for several years before being wooed away to Oregon State University in 1966.

Not all of my traveling was for work. In the summer of 1967, Usha and I went to visit Madan in Kingston, Canada, located about 180 miles from Toronto. I could see that he had done very well for himself—and he had not changed. He was still a wonderful older brother and greatly concerned about me. Madan and his wife, Santosh, took us on a boat ride to visit one of North America's most scenic places, the Thousand Islands. We greatly enjoyed this excursion, which was rich in history and beautiful natural sights. This was just the first of many enchanting warm- weather trips. Over the years, Madan and I would take turns visiting each other during the summer. I greatly treasured these visits and the lifelong friendship I had with Madan.

A Growing Family and Flourishing Career

The fall of 1967 brought with it more great news for Usha and me. Our first son, Ajay, was born in November at Michael Reese Hospital. Now Usha really had her hands full, and I was still spending a lot of time with my students. I began to publish some of their research results and was presenting them at the professional society meetings. Usha noted that I was not spending enough time with our new born son or her. At the time, I dismissed this as unimportant, but looking back, I must confess that she was right.

My initial grant from the NSF ended in 1967, and I prepared to write another proposal. However, this time I focused on the new topic of surface viscosity, which Pintar had begun working on before he left for Michigan Tech. The problem was that Pintar had left for his teaching post without submitting his thesis! I had to sweat for over a year before he handed in his dissertation in the fall of 1968, after which I was finally able to submit my second proposal to the NSF. However, the research findings were well worth the wait. Not only did the NSF fund this initial proposal in 1969, but it continued to fund our work in colloids and interfacial phenomena for the next 40 years.

By 1970, seven of my Ph.D. and ten of my master's students had already graduated, but I still had seven Ph.D. and four master's students working with me. Except for Pintar, who joined academia, the rest eventually all took jobs at private companies. In 1969, Bernie Baker founded his own company, the Energy Research Corporation (now FuelCell Energy), which specialized in battery and fuel cells. Chi joined W. R. Grace & Company in Maryland and eventually served as president of its subsidiary in Seoul, South Korea, before he retired. Punwani joined the IGT and eventually retired as its vice president. Ravi Randhava joined Union Carbide, and ultimately, he and his brother founded their own company.

During this time, a world of opportunity began to open up to me, and I needed to make some difficult decisions about my own career path. In 1969, the College of Engineering at Auburn University contacted me: Would I consider becoming the head of their chemical engineering department? It was certainly an enticing offer. On the advice of Pete Mayfield, my program director at the NSF, I interviewed at Auburn. While the school was quite nice, I realized that I did not want to move to the south. In the spring of 1970, I received an offer of full professorship from the Clarkson College of Technology (Clarkson University) located in upstate New York, and later that year, they approached me to become the department chair. Also, in the mid-1970s, I was approached by the University of Southern California for the department head position and by the University of Houston. I did not pursue any of these positions. My family really liked the Chicago area, and I truly believed I could make a difference by staying at IIT. I felt that IIT offered me the chance to pursue my academic career to my fullest potential. When I first joined the university in 1964, I officially became the first academic of Indian origin in the field of chemical engineering in the entire country. I was incredibly proud to serve at IIT.

The year 1970 was another banner year for me. IIT promoted me to the rank of full professor. Ralph Peck had stepped down as department head after 14 years in 1967. His successor, Bernet S. "Bernie" Swanson, was at the helm when I received this promotion.

In September 1970, Usha and I welcomed our second son, Kern, who was born at Michael Reese Hospital. Now that we had two children, Usha became even busier. However, we were fortunate to be living on a college campus. The diversity of people at IIT made it an ideal place to raise our children. Our boys still fondly remember campus life, especially playing with the many multinational children who were always scampering about.

In addition to my consulting work with both the IGT and the IITRI, I was also consulting for the research department at the Oakbrook-based company Chicago Bridge & Iron. Its research department was located in Plainfield, about an hour-long drive from our campus.

Shortly after Kern was born, we moved to a three-bedroom faculty apartment in Cunningham Hall. We stayed there for the next five years before buying our first house in Westmont. Usha and I were very fortunate that Dharam Punwani and his family as well as

Ravi Randhava's brother, Serge, and his family were also staying in Cunningham Hall, and over the years, they became our extended family.

Although Bob Kintner was well into his retirement, he had not lost his passion for his work. We decided to write an article on IIT's chemical engineering department, which was published in the 1971 summer issue of the American Society of Engineering Education journal, *Chemical Engineering Education*. This was the first time ever that an article written on the history of the chemical engineering department appeared in print.

Vietnam War Era

While the 1960s and 1970s encompassed some of the greatest years of my life, they were also an incredibly turbulent time in the U.S. IIT was not immune from the effects of these events. The university has always had a close relationship with the U.S. military; well-trained armed service members not only make contributions to the government but also to the world around them. Many distinguished members of the armed services have studied at or graduated from IIT. There is no doubt that there were mixed feelings about the war effort.

During the Vietnam War years, there was considerable unrest among students because of the military draft on university campuses nationwide. This unrest exploded in Chicago in 1968 at the Democratic National Convention. Protesters against the war went up against Chicago's finest, some of whom had only recently returned from serving overseas. Massive protests against the war were sweeping the entire country. This unrest intensified after four students were killed and nine wounded during the campus demonstrations at Kent State University in Ohio on May 4, 1970. IIT students also engaged in their own nonviolent protests. There was even a peaceful rally in front of Crown Hall, and most of the classes were cancelled that day. However, I decided to go ahead with my lectures, which seemed to be just fine with my undergraduate classes. After the rally, many IIT students gathered in the auditorium of Hermann Hall where President Rettaliata addressed them. It was a very difficult and tense meeting, and I could see that he was not very comfortable. I remember leaving that meeting thinking that President Rettaliata was not going to continue on much longer at IIT—the atmosphere had changed. Soon after the meeting with the protesters, I believe Rettaliata informed the chair of the IIT Board of Trustees that he intended to leave, and in 1973, he completely stepped down as president.

Life moved on, and the conclusion of the war brought an end to the protests. It seemed like a time to start again. My growing family had definitely decided that it was time for a fresh start. Both Usha and I became naturalized U.S. citizens in 1974. Although Ajay was attending the Lab School at the University of Chicago, we decided that we would not buy a house in the city. Instead, we bought a house in the suburb of Westmont, which is located southwest of Chicago. We stayed in that house for 15 years. Every year, we used to get together with my research group of 10 to 20 master's and doctoral students, as well as postdoctoral fellows, to share good Indian food, great conversation, and spirited volleyball and basketball games in our backyard. Both of our sons enjoyed playing with my students, whom I regarded as my extended family. Needless to say, the students very much appreciated Usha's excellent culinary skills.

With my colleagues in 1965: back row (left to right): Kintner, Swanson, Peck, Wasan, Weinstein; front row: Levenspiel, Fu, Fitzgerald

With Usha at our wedding in 1966

Usha and I coming off the plane at O'Hare in September 1966

President Rettaliata presents me with the Excellence in Teaching award in 1967, with Vice President for Academic Affairs Jim Brophy and Dean of the College of Engineering Ralph Owens applauding (right).

With Ajay in 1969

With Ajay and baby Kern on the IIT Campus in 1970

With my brother Madan and his wife, Santosh, in Kingston, Ontario, in the early 1970s

Chapter 4
Becoming a Leader (1971–88)

Leaders are not born; they are created through trials and experience. Even those who are born into positions of power must struggle to find their own paths. When I was younger, I did not often think about the qualities of a good leader, but like everyone else, I could recognize a great one when I saw one. As the 1970s approached, I found much satisfaction in my life. My tiny family was blooming. I had made some wonderful friends. I was thrilled to be a researcher and instructor at IIT. I was content thoughts of leadership were quite far from my mind. However, I soon found that life had some very big changes in store for me.

Becoming Department Chairman

The year 1971 was the backdrop of a turning point in my career, and there was much to marvel at. Apollo 15 astronauts, who had blasted off for the moon on July 26, returned to *terra firma* on August 7. A new musical, *Grease*, which had made its debut at the Kingston Mines club in Chicago in February, was becoming something of a phenomenon. America was still deeply involved in the Vietnam War. On the home front, there was much discussion at IIT about an increase in the room and board rates for students.

In the summer of that year, College of Engineering Dean Peter Chiarulli asked me to meet with him. I was unsure of what he wanted to discuss but headed over to see him. What met me at his office was a tremendous surprise; he offered me the department head position of the chemical engineering department. I was stunned at this unexpected offer. Initially, I told him I was not interested. I was, after all, content with my research and teaching duties—I was one of the few lucky people to have found his niche in life.

Besides, I knew that Bernie Swanson was doing a great job. Swanson was a true IIT man through and through. His undergraduate, graduate, and doctoral degrees were all awarded to him from IIT. In 1944, he took his M.S. degree at Kansas State University but returned to his alma mater in 1947 to take up his doctoral studies. He became an assistant professor in 1950 and a full professor in 1960, and he had been at the helm of the chemical engineering department since 1967. Swanson was dedicated to improving the education and vocational opportunities for IIT students.

I was stunned by Chiarulli's offer, but perhaps what compounded my surprise was Chiarulli's response—he was dissatisfied with my answer. Without further debate, he indicated that I should think it over and respond to him the next day. As I puzzled through this new turn of events, I decided the best thing to do was to consult someone with excellent insights into my dilemma—Swanson. I invited him to lunch with the idea that he would be the best person to discuss this with.

I asked Swanson point-blank if he was enjoying his job. He replied that he was not; he especially missed teaching undergraduate lab courses. Swanson had served as the department head for almost four years. As we enjoyed our meal, he mentioned that he had been thinking about stepping down and recommending me for the job! At that moment, I informed him about my conversation with the dean earlier that day. Swanson was actually very happy to hear about this. I decided to visit Chiarulli again, this time with an answer I was sure he would appreciate.

Unfortunately, Chiarulli was not as enthusiastic as I had hoped; he had believed our conversation was in strict confidence. However, I went ahead and told him that I would accept the job as department head. I also made two requests: that the position title be changed from department "head" to department "chair" and that I would like him to consult with all my colleagues individually before making my appointment. Chiarulli agreed and was as good as his word. At this time, there were no official policies or procedures in place for senior appointments at the university. This sounds odd now in light of the many different requirements and interviews needed for any departmental position, but the mindset was very different back then. In fact, I had not been asked to submit any dossier for review for my tenure and promotions in either 1967 or 1970.

With the consent of my colleagues and a brand-new title, I began to chair the department in the fall of 1971. I did not lack for excellent advice and guidance. Professor Peck, who had hired me initially, was still in the department, so I had two former heads, Peck and Swanson, to consult with for advice about managing the department.

I was very lucky to become department chair at this point in time. The department was full of incredible instructors who were dedicated to education. Ralph Peck, in particular, was an outstanding educator. I felt he should have been recognized with the university-wide Excellence in Teaching award. I was happy to nominate him for this honor, which he then received in 1973. In the following years, I also nominated Professor Swanson and Professor William Langdon for this award, which they both received and greatly deserved.

Modernizing the Curriculum and Managing the Budget

The chemical engineering department had an excellent reputation in teaching and preparing students for professional careers. However, as the department chair, I needed to enhance departmental research activities and recruit new faculty to modernize the curriculum. After gaining approval from the dean, I managed to hire Jim Vrentas in 1972 to join our faculty as an assistant professor. He was responsible for introducing a new course on transport phenomena. This course was already being taught at many leading chemical engineering departments across the country. We really needed to offer it in order to modernize our curriculum. Vrentas also pursued his research in polymer processing. He had worked with Larry Duda at Dow. Duda had joined Penn State University as an associate professor the year Vrentas joined us, and together they had published a number of excellent papers on diffusion in polymers before joining academia.

After Vrentas joined IIT, he and Duda jointly submitted proposals to the NSF, which were funded for a number of years. (After becoming a full professor, Vrentas left IIT and joined Duda at Penn State in 1980.)

In January 1972, at the beginning of my second semester as chairman, I learned to my great surprise that the department had overspent its expense budget; there were no funds left to buy chemicals and supplies for the undergraduate labs. I discussed this problem with both former chairs, Peck and Swanson, and they both said, "Join the club." Not only were they aware of this, they told me that they would routinely overspend the supply budget, and the college dean would bail the department out at the year's end. Running deficits did not appeal to me, so I began to raise funds for the department from industry. I first wrote to request a donation from the CEO of Chicago Bridge & Iron, a company for which I used to be a consultant. Within a month of the request, I was surprised and grateful to receive a check for $2,500. I took the donation to Dean Chiarulli and explained to him what it was for and left it with him to deposit into our department account. After two weeks, he informed me that the department could only use $1,250; the remaining funds would go into the general fund of IIT! Upon further inquiry, I learned that donations could be fully used by the department only if the funds were earmarked for restricted purposes. I suddenly understood that this was the financial reality of IIT; the university needed money too. Consequently, when soliciting extramural funds, I made sure to request that the donors restrict their contributions to the use of the department. These were very big lessons for me as a first-year chair.

Building Alliances with Industry

One of my first initiatives was to strengthen the department's ties with the industrial community. In 1972, I re-established the Industrial Advisory Committee, which had been created in 1969 but had disbanded. The new committee was comprised of almost a dozen members, half of whom were prominent departmental alumni and the rest of whom were from companies which had not hired many graduates from our department. All the members of the committee held leadership positions in their organizations. As a result, we were able to raise some unrestricted funds to support the department from within industry. Perhaps even more importantly, through the activities of this committee, there was constant and continued improvement in communication with businesses. This directly enriched the instruction in several of the courses offered by the department. For example, two practicing process design engineers from Universal Oil Products (now Honeywell UOP) teamed up with Professor Langdon in the department to teach the capstone process design course to our senior classes.

Food Technology

The committee was just the beginning of major changes in the department. I recognized early on a particular need of the local food processing industry: the ability to recruit chemical engineers with a strong knowledge of food technology. However, this was an area of significant challenge; after all, there were *no* academic institutions in the Chicago area offering this expertise. We established a new certificate program in our department to meet this industry need. The Proficiency in Food Technology certificate was comprised of six courses taught by adjunct instructors recruited from the food industry. The instructors were selected based on the recommendations of the Education Committee of the local section of the Institute of Food Technologists (IFT). This section was chaired by Elwood Caldwell, vice president for research at Quaker Oats Company. The courses addressed key topics such as food process engineering, food packaging, food chemistry, and nutrition. Students flocked to these classes; both full-time students and professionals who worked at local companies and attended classes part-time found them useful. Upon the completion of the program, students were presented with their certificates at the annual meeting of the IFT in Chicago, which reinforced the value of their educational experience.

The demand for food technology coursework was strong. Around this time, I began working closely with Professor Art Siedler, head of the food science and human nutrition department at the University of Illinois at Urbana-Champaign, as that department was already offering a master's degree program via their extension program in Chicago. Instructors from Urbana were sent to teach classes at the University of Illinois at Chicago campus. We worked out an agreement that the University of Illinois master's students studying at the Chicago campus could take our food technology courses and then simply transfer these classes to the University of Illinois for credit. This arrangement avoided any duplication of offerings in the Chicago area by two competing institutions. Caldwell was most helpful in facilitating this arrangement. However, the seeds of a much greater endeavor were planted through this partnership, but it would take almost two decades to come to fruition. In 1988, my collaboration with Art Siedler and his food science faculty at the University of Illinois resulted in the establishment of an incredible institution—the FDA-funded National Center for Food Safety and Technology at IIT.

The Energy Crisis

Food technology and safety were foremost on my mind when politics took control of a key area of engineering. Starting in 1973, the Organization of Petroleum Exporting Countries (OPEC) made a startling move, imposing an oil embargo on the U.S. This unexpected turn of events had a devastating impact on the American economy and resulted in an energy crisis. National players were taking action to address this crisis, and as the department chair, I knew I needed to confront this problem head-on. I began to strategize by hiring new faculty in vital energy areas, and I welcomed the opportunity to work closely with the two neighboring institutes at IIT, the Institute of Gas Technology (IGT) and the IIT Research Institute (IITRI), both of which had substantial involvement in energy-related research activities. In 1975, I recruited Rob Selman, who was working at Argonne National Laboratory (ANL), to join the department. He had received his Ph.D. from University of California, Berkeley, and his area of study was electrochemical engineering with a specialty in batteries and fuel cells. I introduced him to the research group working in a similar area at the IGT, where he was hired as a consultant very shortly after he joined the department. Selman also continued his interactions with his former colleagues at ANL and established cooperative research activities with both the IGT and ANL. Selman had a stellar career at IIT and was responsible for establishing the first national research center in this field, which was funded by the U.S. Army with an initial grant of $2 million. (Selman now serves as an emeritus professor in the department. Since his area of specialty is so fundamental to the progress in energy studies, the department recruited a new faculty member upon his retirement.)

I took a multipronged approach to addressing the energy issue; supporting electrochemical engineering was just one part. An important piece of the energy puzzle is oil. America is a motor nation, and few images from the time of the embargo were as stark as those of vehicles backed up for miles outside of gas stations, waiting for their ration of fuel. In the summer of 1974, I secured a seed grant of $1,500 from the Graduate College at IIT to support one of my graduate students and initiate research in enhanced oil recovery (EOR). After the summer was over, I submitted a proposal on this topic to the American Petroleum Institute, which provided a grant of approximately $30,000 to continue the work.

Research involves considerable expense and obtaining resources for that research has always been a challenge. However, greater funding opportunities arose in EOR in a variety of energy-related research areas after the embargo. The NSF established a separate program, the Research Applied to National Needs (RANN), which operated from 1969 to 1977, to specifically address issues that may affect the entire country. I submitted my proposal on EOR to RANN and received funding in the amount of $200,000. The federal government, under the leadership of then President Jimmy Carter (an engineer), also established another new agency, the Energy Research and Development Administration (ERDA), which started coordinating all energy-related research funded by all the federal agencies. Upon the completion of my grant from RANN in 1978, I applied to the ERDA (now the Department of Energy) and received $450,000 for the continuation of my EOR research.

Establishing the Industrial Research Advisory Committee

Any researcher must ensure that his or her work is relevant to others. Engineering researchers do not simply pursue their areas of interest; they carefully consider the impact and practical applications of their work. I, too, was concerned about the practical outcome of my research. With this in mind, I set up the Industrial Research Advisory Committee comprised of the chief scientists and research directors from about 20 global oil and chemical companies. Their input into my research direction was invaluable, and I gained much from their insights. I was especially concerned about educating our students so they could obtain employment at these types of firms. We were then conducting fundamental research aimed at investigating the underlying mechanisms for EOR based on chemical flooding methods.

There have been some surprises in my career, but I think one of the most amazing occurred after the creation of the Industrial Research Advisory Committee. Most of the companies that participated in this advisory committee found our research interesting. As a matter of fact, they found it so useful that some of the companies started sending us gifts in amounts ranging from $5,000 to $15,000 every year to support our research. They also quickly began recruiting graduates from my lab to work for them.

While the extra money was certainly a tremendous gift, I knew that it had to be managed carefully in order to derive the most value from it. I decided to hold some of the gift money in reserve, as I had sufficient extramural federal funding to support all of my graduate students— approximately 12 to 15 students at any given time. By 1986, I had saved approximately $600,000 of the industrial gift funds. With the assistance of the Office of the IIT Comptroller, this surplus was deposited into a quasi-endowment account labeled "EOR Project." As the principal investigator of the funds, I decided to use them to support Research Professor Alexander D. Nikolov and some of the doctoral students who worked with me.

Recruiting Faculty Talent

At the beginning of the fall of 1977, I recruited Professor Dimitri Gidaspow to work with me, as he was considering leaving the IGT. Gidaspow, who was from the Ukraine, had earned his master's in chemical engineering from the Polytechnic Institute of Brooklyn in 1960 and his Ph.D. from IIT in 1962. Ralph Peck had recruited Gidaspow as a teacher immediately upon his graduation; this was again Peck's remarkable ability to understand and recruit the finest in scientific talent.

I had asked Gidaspow to work with me to further strengthen the energy-related research activities in the department. I was so pleased that he agreed, since he had a quick mind and a deep understanding of some of the most complex topics in chemical engineering. Gidaspow's research interests were in the areas of multiphase flow and particulate processing with particular industrial applications related to research in gas and oil. He and I would jointly supervise several doctoral students working in energy-related research over the next 10 years. Our collaboration created a unique opportunity in our research. For example, we secured a grant from Amoco to remove very fine particles from coal-derived liquids to produce clean fuels. During our partnership, we also discovered that the coal-derived liquids contained micron- and submicron- sized particles that were charged due to the adsorption of the asphaltic compounds on their surfaces. We worked together to develop methods to separate them from the coal-derived liquids by using an electric field. In 1980 and 1981, we obtained two patents based on our novel method of crossflow electro filtration. The NSF was willing to support our research on this topic, since it showed great promise. In 1986, Gidaspow and I received the NSF's Special Creativity Award for our research on the separation of particles from nonaqueous media based on electro kinetic phenomena. The people who make up your research team and work environment have a strong impact on your success. It is not always easy to recruit scientific talent to academia. Because of the oil crisis, industries were hiring many doctoral graduates. It was a very difficult time for academic departments to recruit engineering faculty. The Amoco Foundation, under the directorship of Don Schroeder, a chemical engineer, set up a special fund to help academia recruit new faculty in chemical engineering. I was able to secure startup funds from the Amoco Foundation to pay for the first three years of the three-month summer salary for new faculty member Dimitris Hatziavramidis. A former doctoral student of Tom Hanratty, Hatziavramidis had done his postdoctoral research at Cornell University. He replaced Jim Vrentas, who had left to join Penn State University.

Our department has never lacked for talented, pioneering educators with a sincere commitment to their work. For example,

Henry R. Linden, a member of the National Academy of Engineering, was appointed a research professor in the department in 1978. When Linden retired as the president of the Gas Research Institute in 1986, I asked him to join the department full- time as the Gunsaulus Chair Professor. At the time, Linden was well known nationally as the true "energy czar" (in addition to being quite fond of downhill skiing). Hamid Arastoopour, one of Gidaspow's Ph.D. students, was another researcher working at the IGT who joined us in the 1980s. Selim Senkan from MIT, with an interest in combustion, rounded out our team. Our department became a powerhouse in energy-related research in the Midwest.

We were not only performing fundamental research and educating students, we were innovating in research and academics. All of the members of our department felt that what was taught in the classroom should be relevant to industry. For example, Linden and Arastoopour introduced a new program called Energy, Environment, and Economics, which highlighted the important bonds between energy use and the world around us. Arastoopour is now serving as the director of IIT's Wanger Institute for Sustainable Energy Research.

Becoming an Interim Dean

When the chemical engineering department began growing, I found another opportunity opened up to me. In the fall of 1977, IIT President Thomas L. Martin Jr. and Provost Sidney Guralnick recruited me to serve as the interim dean of the Armour College of Engineering when Dean Chiarulli suddenly stepped down. I, in turn, recruited Gerald "Jerry" Saletta from the electrical engineering department to serve as my associate dean on a half-time basis, as he also kept teaching. At the time, the dean's office had one very able secretary, Ms. Sitgraves. During that year, I made some administrative changes, including naming a new interim chair in the mechanical, material, and aerospace engineering department. During that year, Saletta and I primarily focused our efforts on developing a reward and incentive system for the faculty to both recognize and increase their productivity. Upon the completion of my term as interim dean for the year, I personally recommended Andre Vacroux, chair of the electrical engineering department, for the position of dean. Vacroux served in that position for eight years until 1987, when Meyer Feldberg was appointed the new president of IIT.

Expanding to the Health Sector

When I resumed my duty as chair of the chemical engineering department in 1978, I began to focus on developing a second strategic area of research for the department: health. In 1980, IIT established the Pritzker Institute of Medical Engineering with an initial endowment of $5 million from one of our generous alumni, Robert A. Pritzker. Robert Arzbaecher, a professor of electrical engineering, was appointed to be the institute's director. In consultation with Arzbaecher, I recruited Richard L. Beissinger, a chemical engineer, as an assistant professor in the chemical engineering department.

Beissinger had done his doctoral thesis work on a biomedical engineering project with Professor Edward F. Leonard of the chemical engineering department at Columbia University. His postdoctoral work was in blood rheology with Professor Michael C. Williams at the University of California, Berkeley. Arzbaecher provided the funds to support Beissinger's research. Beissinger was also successful in getting some funding from the National Institutes of Health (NIH). Over the years, Beissinger and I collaborated on a research project on artificial blood based on a double emulsion formulation, which required my expertise in colloid and interfacial science. We jointly supervised one doctoral and two master's students on this project. Ultimately, we obtained two U.S. patents and co-authored several papers on it.

In the early '80s, I also recruited three new faculty members, including Satish J. Parulekar (who is still on our faculty today) to conduct research work in biotechnology. I was able to secure funding in the amount of $300,000 from the Amoco Foundation to support our faculty's initial research efforts in biotechnology. This research team was housed in the Life Sciences Building (now the Robert A. Pritzker Science Center, dedicated in 2016) so that our departmental faculty could collaborate with the faculty of the biology department, which was also located in that building. In 1988, I convinced the senior research management of Corn Products International (now Ingredion Inc.), whose vice president, George Hoff, was our departmental alumnus and a member of the Industrial Research Advisory Committee, to loan us their most senior biotechnologist, Fred Keller, to work with our faculty on our campus for a year. Not only did Keller join our faculty, but he also brought some equipment with him to IIT, and his full salary was paid by his company. And, I recruited Ali Cinar from the University of Wisconsin-Madison. Cinar worked in the process control and system engineering area, and he is now director of IIT's Engineering Center for Diabetes Research and Education, which is funded by the

NIH. Previously, Cinar served as the dean of the Graduate College and vice provost for the Office of Sponsored Research and Programs at IIT.

Polymer Science and Engineering

In addition to establishing the two strategic areas of research energy and health, a new opportunity arose to build a third area of research. Amoco purchased a plastics products' company, and so there was a need for experts in polymer science and technology. Plastics are a ubiquitous part of daily life and reach deeply into industry and business. There was definitely a strong need to educate graduates in polymer and material processing, and, at the time, no such program was being offered by academia in the Chicago area. I decided to approach the leadership about establishing a center for polymer science and engineering at IIT. At the time, Robert D. Cadieux, an IIT trustee, was Amoco Chemical Company president. It was early in 1987 when Henry Linden and I went to discuss this proposal with Amoco. I requested seed funding of $500,000 from the foundation. Fortunately, we found both Cadieux and the director of the foundation to be receptive to our proposal. After I resumed the position of interim dean of the engineering college in 1988, I asked Linden to prepare and submit our proposal to Amoco for possible funding. The Amoco Foundation ultimately approved the funding, and the Center for Excellence in Polymer Science and Engineering was established in 1990, under the direction of the chemical engineering department at IIT. The center developed Master of Science degree programs in chemical engineering, metallurgical and material engineering, and chemistry with a specialization in polymer science and engineering. Additional funding to support this center in the amount of $300,000 was provided through a State of Illinois Technology Grant.

Doctoral Student Excellence

Meanwhile, my own funding to support my EOR research grew in the 1980s; I had received a major grant from the ERDA and a continuing grant from the NSF, as well as funding from several companies. In addition to my administrative duties as department chairman, I had about 22 graduate students working with me.

During this period, my research group included some of the best doctoral students I have ever had. For example, David A. Edwards, currently the Gordon McKay Professor of the Practice of Idea Translation in the John A. Paulson School of Engineering and Applied Sciences at Harvard University and a member of the National Academy of Engineering, authored the textbook *Interfacial Transport Processes and Rheology* with Howard Brenner and me. Rama Ramakrishnan, also an outstanding student, now serves as a scientific advisor and head of the CO2 Mitigation department at Schlumberger-Doll. Raju Borwankar, who died at age 52 from sarcoma, became the director of analytical sciences for North America at Kraft Foods. Jeffery P. Perl is president and principal engineer at Chicago Chem Consultants Corporation, a company that provides innovative environmental and chemical engineering services. Venu Menon is vice president of technology development at Texas Instruments. Daniel Huang went on to serve as a senior fellow and associate director at Novartis Pharmaceuticals. Li Ting has pursued otherprofessional goals. Alex Nikolov, who was initially supported by the extra funds provided by industry leaders, also began working with me at this time. His expertise in colloid and surface chemistry is unparalleled. Nikolov joined my group as a senior research associate and is now a research professor in the chemical engineering department. Nikolov graduated from the University of Sofia in Bulgaria, where we first met in 1979.

A Thriving Chemical Engineering Department

By the mid-1980s, both Ralph Peck and Bernie Swanson had retired, leaving a total of eight faculty members in the department. (In honor of Peck, the department established the annual Ralph Peck Memorial Lecture Series, which continues to this day.) Research productivity of the chemical engineering department faculty was at an all-time high; with nearly one-third of the academic-year salaries for eight faculty members charged to their extramural grants. Forty- five full-time doctoral students in the department served as research assistants, with their full tuition paid to IIT by the faculty research grants. The department ranked among the top 20 departments in the country, as measured by an average of over $150,000 in extramural grant dollars per faculty member.

During the late '80s, after a consultation with one of my former students who was then vice president of manufacturing at Abbott Labs, I also explored the possibility of offering a graduate program in pharmaceutical engineering in the chemical engineering department. In 1987, as I was about to resume serving as interim dean of the college, I advised my successor, Linden, to undertake this initiative. While the department offered a few courses in pharmaceutical engineering, a full program was not fully developed at this time. However, I am pleased to report that as I write this in 2018, Natacha DePaola, who is the Carol and Ed Kaplan Armour College Dean of Engineering, has made this one of her new initiatives.

As chairman, I managed the finances of the department based on a simple formula. The total expense budget of the department including salaries was not to exceed one-third of its total revenue, including the net tuition fees paid by the students plus the indirect cost charged to the extramural grants secured by the faculty. I found this formula served us well in keeping the department in good fiscal condition for those 16 years of service. The funds that I raised from industry for unrestricted use in the department were put aside in a separate account and used for new faculty startup funds and to support research assistants when their advisors' grants ran out of funds to support them. This normally happened when the faculty had applied for a renewal of the grants but had not yet heard back from the funding agencies. These funds were also used to support faculty trips to attend professional conferences overseas and send graduate students to present their research results at professional society meetings. These funds grew under the departmental leadership of both of my successors, Henry Linden and Hamid Arastoopour.

After Meyer Feldberg became president in 1987, he asked me to chair the search committee to recruit a provost. We were successful in recruiting Mel Bernstein, a metallurgist from Carnegie Mellon University (CMU). After Bernstein joined us in 1987, both he and Feldberg asked me to serve as dean of the College of Engineering, which I agreed to for one semester beginning in January 1988. I appointed Henry Linden as interim chair of the department during that period. I stayed in that position until the end of the summer, when the new dean, Richard Saeks, an electrical engineer and department head at Texas Tech, was recruited to join IIT.

As the 1980s drew to a close, I found myself a changed man. I had begun my educational service with the intent of teaching students and revealing the beauty of natural processes. What I discovered through the various leadership positions that I held made a powerful impact on my life. I realized the importance of leadership and saw firsthand how my decisions influenced my own work and that of others. I was particularly captivated by the way management issues played out in the academic lives of our students. While I had not sought out a position of leadership, I found that I had grown significantly from my experiences. This knowledge laid the foundation for taking on even greater challenges.

President Martin accepting a check from a representative from CPC Industry in 1979

With Dimitri Gidaspow in 1982

On a burro at Santorini Island in 1984

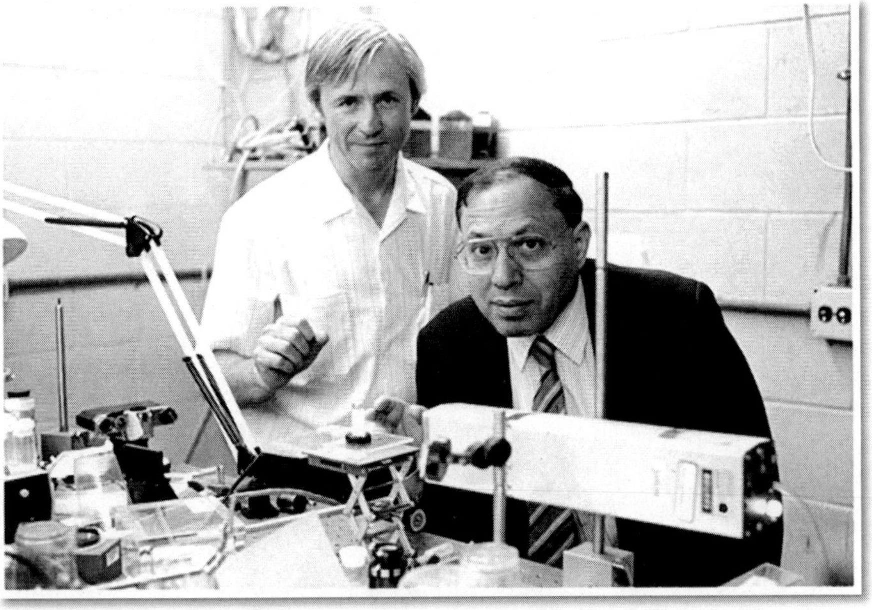

With Alex Nikolov in 1987

Chapter 5
The Birth of NCFST and IFSH

In December 1987, Fred R. Shank, director of the FDA's Center for Food Safety and Applied Nutrition (CFSAN), approached me when I was still department chair. The FDA had been at the forefront of protecting the nation's food safety, but, in 1985, outbreaks of food-borne pathogens caused it and the Centers for Disease Control and Prevention (CDC) to take a more proactive role in addressing the public's needs. Part of that approach included increasing the collaboration between academic and private entities and using their talents and facilities to fight against food contaminants.

When Shank called me, I was more than a little curious, as it was not often that I got a call from the director of the nation's vanguard on food and drug safety. I asked why he had chosen to contact me; he mentioned that the Department of Chemical Engineering at IIT was the only academic department in the Chicago area that offered a food technology program. He asked me if I could suggest to him a suitable lab space with a pilot plant facility in the Chicago area; he was interested in relocating two dozen of the FDA scientists from their antiquated research laboratories in Cincinnati.

After giving this some thought, I told him that CPC International (now Ingredion and still making KARO syrup, SKIPPY peanut butter, and Hellmann's Real Mayonnaise) had closed their research facilities at their Moffett Technical Center in Bedford Park, a suburb southwest of Chicago. I advised him to contact the former president of CPC's Corn Wet Milling Division, George

E. Hoff, an IIT alumnus (CHE, M.S. 1950) and a member of our department's Industrial Advisory Committee since the early 1970s.

After speaking with Hoff, Shank called me back to let me know that CPC was intending to sell its research facility, rather than rent it out. Although this news was somewhat disconcerting to me, it actually served as the momentum for the next and more important part of our conversation. Shank and I then began to discuss the idea of creating what would become the National Center for Food Safety and Technology (NCFST).

Shared Vision

Shank and I shared a vision of establishing a unique consortium of government, academic, and industry leaders, working together toward the goal of improving food safety. To his credit, Shank recognized that no one sector of the food safety community could afford to conduct all the food safety research required in order to keep pace with the rapid technological developments and evolving threats to public safety. We both envisioned a collaborative center where common food safety issues could be addressed in a neutral environment.

My conversation with Shank sent my mind into a new direction; I began to wonder how a national center might actually come to life in Chicago. CPC wanted to sell the Moffett Technical Center, but it was unlikely that IIT could afford the nearly $7 million price tag. How could I possibly find another location like this one? After all, the Moffett Center was not simply a collection of buildings and labs; in its heyday, it had been one of the most advanced food laboratory facilities in the nation. After puzzling it over, I decided that there was only one thing to do—take the direct approach.

With my mind full of dreams for a new center, I reached out to George Hoff and explained to him what could be thought of as nothing less than a gamble. I asked if CPC would simply donate the land and the buildings to IIT. Hoff, an IIT veteran, was no stranger to the financial struggles faced by researchers in academia. He immediately discussed it with the company's CEO James R. Eiszner (a University of Illinois alumnus). Eiszner liked the idea that the food industry would thus have an opportunity to work with the FDA in an academic and neutral setting; he saw great benefits to the food industry by creating such an entity.

After receiving this encouragement from Eiszner, we reached the stage that many researchers dread—the paperwork. However, the paperwork for this project did not seem to be too tedious. Clyde Watkins, vice president of external affairs at IIT, and I both worked with Hoff to prepare the necessary documentation for the company to donate the property to us.

The Moffett Technical Center

In March 1988, IIT was given the Moffett Technical Center, a $6.9 million gift. Designed by Schmidt, Garden and Erickson in 1947, this newest addition was a veritable marriage of art and science. The center was named in honor of George M. Moffett, one of the presidents of Corn Products before it became CPC International. The bas-reliefs on the main building were designed by Lee Lawrie, as was the famous Argo section, showing engineers and researchers hard at work. At the time, the center was comprised of five buildings, including the research laboratory (Building 90) and the pilot plant (Building 91). The research laboratories occupied the second and third floors of the building; the first floor had a reception room, a mail room, offices, a kitchen, a cafeteria, and a library. More importantly, there were heavy-duty physical laboratories and library stacks in the basement, making the building an effective springboard for research. Its sub- basement contained a machine shop, along with the usual mechanical services and storage areas. The manufacturing laboratory in the pilot plant building was another gem. The front section had four floors, with the first two housing a machine shop, locker rooms, offices, and conference rooms. Adjacent to this facility was the company's small plant for manufacturing special chemical products, including processed corn such as sweeteners and various other consumer products.

While I was basking in the warm glow from the gift of the Moffett Center, there was a minor nagging problem. It began with a phone call from Shank about moving his researchers. In the excitement of starting the center, I had failed to do something very important: I did not yet have the approval of the FDA to do so. At the time we secured the gift of the center, it was my understanding that IIT would submit a sole-source proposal to the FDA to fund the NCFST and also that the FDA would rent space from IIT to relocate their scientists and staff. Since we had the center, I thought all was well; however, on May 2, 1988, a month prior to the due date of IIT's proposal to the FDA, the FDA issued a Request for Proposal Application (RPA) in the federal register.

Our Proposal to FDA

Because the request appeared to open the door to other bidders as well, I suddenly became concerned about the future of the center. Shank explained to me that an open competition was necessary for the agency to award a federal grant. This put a lot of stress on me; suddenly I had to prepare a winning proposal. Imagine how I felt!

I had raced toward the dream of having a national center in which new food processing technology and inspection strategies could be developed, and I had secured an ideal research facility from CPC, but I now faced a potential crisis. I was especially troubled, because I had asked for the Moffett Center for the sole purpose of establishing a collaborative facility for research in food safety. I had only one month to write the proposal, while competing with almost 40 well- established food science academic departments, some of which already operated at land-grant colleges. I hardly slept that month. I think that the excessive stress is what eventually landed me in the hospital for emergency triple-bypass heart surgery.

I realized that to write a truly great proposal that would make IIT the clear winner in this bidding contest, I needed to interview industry leaders and learn how they would like to interact with the FDA in a collaborative environment. Fortunately, one of IIT's strengths has always been its strong ties with industry, and Chicago has been home to many renowned and innovative companies in the food industry where notable advances have been made in food safety. In 1909, for example, the city of Chicago required milk pasteurization and was the first U.S. city to do so. As a scientist and engineer, I knew it was vital for the food industry to explore emerging areas of the technology that the new center would concentrate on. I wanted to hear firsthand what they needed.

I reached out to the vice presidents of research at Kraft Foods and Quaker Oats, two of Chicago's most iconic food businesses. I had recruited two of my colleagues to help me prepare the proposal: Richard H. "Dick" Snow, head of the chemical engineering group at the IITRI and an IIT alumnus who had done his doctoral work with Peck, and Irwin "Irv" Fieldhouse, another alumnus who, in 1936, was one of the first employees at the IITRI's predecessor institution, the Armour Research Foundation.

The three of us met with the two industry leaders at their offices. Later, Snow, Fieldhouse, and I hunkered down to draft our collective thoughts, carefully considering everything we had heard. Both vice presidents had given us important insights into their businesses. I had them review the draft to ensure that we had accurately captured their recommendations. Then, I also solicited suggestions and comments from a few other industry experts.

I was fortunate to have a strong team and sounding board; many of these experts had also been part-time course instructors in IIT's certificate in food technology program. Their input greatly strengthened our proposal, but perhaps the *pièce de résistance* were the testimonials we included in the proposal: encouraging statements from the chairs and CEOs of several companies who expressed their support for the establishment of the center. In addition to Kraft and Quaker Oats, other supportive companies included FMC Corporation, Wm. Wrigley Jr. Company, CPC, and Dean Foods. I was also able to persuade Art Siedler, head of the food science and human nutrition department at the University of Illinois at Urbana-Champaign, to be a cooperating member of the center.

In our funding requests in the proposal, we included details of the research laboratory and pilot plant facilities of the Moffett Center. An initial request of $700,000 was for the first year; the total request was for $3.7 million over the first three years of operation. The IIT proposal, along with others from various schools, went through an external dual-review process. We had to wait three months for the FDA to render its decision. Finally, we received the good news—the FDA had selected IIT's proposal with funding to start on September 29, 1988. IIT would be granted $700,000 for operating expenses for the first year, $1.2 million for the second year, and $1.8 million for the third year.

Time for Action

The long wait was over; it was now time for action. The initial government funding was vital to the center's existence, but it would not cover the cost of all the necessary work. I reached out to Raymond C. "Ray" Tower, president and COO of FMC Corporation and a member of the IIT Board of Trustees, to pledge $100,000 for the first industry membership in the center. After securing his pledge that morning, I asked him if he could set up meetings for me with the chairs of Kraft Foods and Wrigley. Tower acted quickly; I was in the Kraft and Wrigley offices that very same day, asking for their support, and Tower was with me when the CEOs of the two companies generously made pledges of $250,000 and $100,000, respectively. In one whirlwind day, we had garnered a total of $450,000 in memberships.

But this was only the beginning. The initial investments were vital in funding the center, but equally important were the structure and the nature of the center's activities. As with any new project, there were several questions to be answered: How did businesses view the center? What roles did they want to play? How should the FDA and IIT be involved? What were the roles of the University of Illinois and the IITRI as cooperating partners?

Shank and I then embarked on a series of friendly fact-finding meetings with businesses, seeking their input on how the NCFST should be structured and run, the kinds of research the center should conduct, and the roles that the FDA, industry, and IIT should play in the NCFST. The results were incorporated into the NCFST's burgeoning development.

With new ideas for operating the NCFST and a vision for its potential research areas, the center sprang to life. The University of Illinois at Urbana and the IITRI had participated in the consortium from its very beginning. With the help of Shank, I quickly recruited other industry leaders as well, such as the BSN Group, Dean Foods, McDonald's, Nabisco, and Pfizer.

I managed the NCFST until a new director was recruited. Following a national search in 1989, Richard V. "Dick" Lechowich became the first director of the center. He had worked in industry at General Foods and ABC Company and in academia, serving as department head of the food science department at Virginia Tech. Isabel D. Wolf, his wife, was appointed the director of communications and marketing of the center. She had served at the

United States Department of Agriculture and later in the extension education program of the food science department at the University of Minnesota.

The positions at the NCFST began to fill swiftly as the buildings were rehabbed and renovated to adapt to their new purposes. In September 1989, Shank transferred about two dozen FDA scientists from their Cincinnati research laboratories to Chicago. During the renovation, the FDA team that was to become the food engineering branch and the quality assurance branch of the center was shuffled over to the IITRI. In addition to renovating the center's laboratory space, a new elevator was installed, bringing the total cost of the renovation to about $800,000. (It was paid back with 10 percent interest to IIT by the FDA over the next 10 years.)

David J. Armstrong, a long-term FDA employee, was appointed associate research director of the FDA located at the NCFST. By July 1990, all FDA scientists and staff had moved into the renovated offices and laboratories in Building 90 of IIT's new Moffett Campus. Robert Opila, a veteran chemical engineer and IIT alumnus, joined the NCFST as manager of the pilot plants. He had designed food processing systems and facilities for various international food companies all over the world.

The NCFST was dedicated on May 22, 1991. David Kessler, FDA commissioner, performed the ribbon cutting at the ceremony and spoke eloquently of the potential of the NCFST as a national resource for food science and technology. Other speakers expressed their thoughts and high hopes, including Howard M. Dean, chairman and CEO of Dean Foods; Robert G. McVicker, senior vice president of technology, quality assurance, and scientific relations at Kraft Foods; Donald E. McElmury, vice president of operations and technical support at CPC; Shank; and IIT President Lewis M. Collens. The event was covered by the local media and the Associated Press, which reported on the NCFST's grand opening to the entire nation.

In the fall of 1991, we initiated the NCFST's new master's program in food safety and technology, designed for individuals with degrees in food science, engineering and/or the biological and physical sciences who wanted to become food safety experts. It was first offered in the evenings and later expanded as its popularity grew. The courses were team-taught together with senior scientists of the FDA.

The NCFST Library was established in 1993 by merging materials loaned by Corn Products, the FDA, and the IIT libraries into the national center's library. The total collection originally contained approximately 3,000 books and 369 journal titles. Today, it is one of the best food science libraries in the nation and also operates as a depository for materials from the United Nations' Food and Agriculture Organization (FAO). In 1993, the NCFST built a $1.2 million biotechnology pilot plant, made possible with a

$540,000 Technology Challenge Grant that I secured from the Illinois Department of Commerce and Community Affairs (DCCA); the equipment was purchased with FDA funds. The plant included three fermenters, computer-linked sensors and controls, and a centrifuge for harvesting cells. This major fermentation equipment was augmented by downstream ultrafiltration and reverse- osmosis processing systems. The collaborative research program began at the end of 1990.

Industry partners helped determine the research priorities at the NCFST, through a well- defined process (see figure 5-1 for the

NCFST project selection timeline). Research planning depended heavily on review by the Technical Advisory Committee (TAC) and its steering committee. They were assembled by David Armstrong. Four working groups were formed by the TAC: Safety Assurance, Food Processing, Food Packaging, and Biotechnology. Participants were drawn from the founding and affiliate members of the NCFST. Jonathan W. DeVries of General Mills served as the long-term chair of the TAC. In 1993, each of the four working groups held their first meetings to define their specific research area priorities, and they recommended approval or renewal of specific projects. About 30 researchers attended each of these meetings.

One of the major benefits of the NCFST was the opening of communication channels between industry and FDA representatives who participated in the working groups. According to Armstrong, the industry feedback that they were able to provide to the FDA had a significant impact on the food safety policy established during the previous year.

Robert McVicker of Kraft Foods served as the first chairman of the NCFST's Oversight Advisory Committee (OAC). In that position, he helped recruit fellow food industry members and especially encouraged the participation of Kraft researchers in its working groups. His successor was Charles Feldberg, vice president of health, safety, and quality assurance at CPC. This committee was convened twice a year by Shank and me.

Over the years, the NCFST staff has grown and changed. In addition to its founding members, about 15 new affiliate members and some 30 associate members joined. The FDA also housed the Division of Food Processing and Technology (later renamed the

Division of Food Processing Science and Technology) in a nearly 30,000 square-foot special research facility in Building 90. Today, approximately 50 FDA research scientists and staff members work in this division, with Richard E. McDonald serving as the division director, and about 100 graduate students also work in the center.

After leading the NCFST to its most successful year, Dick Lechowich and Isabel Wolf retired in the spring of 1998. Charles Sizer, the former general manager of the collaborative center's member company Tetra Pak, was appointed the new director of the NCFST. He had served as a faculty member at Iowa State University and had worked at Frito-Lay Research. He was a fine choice to succeed Lechowich. To start with, Sizer had secured a donation from Tetra Pak of an aseptic particle processing system, then valued at $1.2 million. He was a scientific powerhouse in his own right, having secured over 40 patents and more than $22 million in grants before he moved on to another position in 2003.

The NCFST continued to blossom through its many supporters, and in 1999, the FDA renewed the center's cooperative agreement. This renewal was part of the agency's recognition that the work at the NCFST complemented and fed into the Food Safety Initiative (FSI) Risk Assessment. The FDA also paid the NCFST a compliment; the FDA had established a similar type of collaborative group at the University of Maryland, based on IIT's successful NCFST model. The NCFST's work helped support the activities at the newer facility, the Joint Institute for Food Safety and Applied Nutrition (JIFSAN). I served on the search committee for its director. Then, the FDA funding of the NCFST grew from $2.0 million annually to $3.0 million. For more than 15 years, Sizer,

Armstrong, and I regularly interacted with Karen Carson, the FDA project officer. We much appreciated her cooperation and guidance in the successful management of the center.

Our Pride in NCFST

I was very proud of our work at the NCFST, not only because I had seen the center through its nascent stages, but also because it operated on principles that were important to me. Science can improve our lives, help others, and make the world a much better place; however, the most advanced science is incomplete without the collaboration and teamwork of dedicated, caring people. There are many other entities that share this approach to research and collaboration.

In the spring of 1999, I received a great compliment from Jane Henney, MD, the new FDA commissioner. During her visit to the NCFST, she said to me: "NCFST is a living example of what I am talking about, both in terms of being able to make our decisions on a science base and also looking for opportunity in which we can leverage our own intellectual capital with that of others." The first woman to serve as an FDA commissioner (appointed by President Clinton), Henney had been trained to become an oncologist and served at the National Cancer Institute—thus she was no stranger to applying science to help others. Her kind words were not only an acknowledgment of the hard work we were doing at the NCFST, but also of my constant desire to serve others.

Since the NCFST was now operating so well, it seemed that a local version would be in order. While Kraft and other large companies had a national and international impact, smaller businesses could definitely benefit from the type of research being carried out in their home state. Following this thought, Irv Fieldhouse and I fleshed out a proposal to the state about a new consortium that would offer a competitive edge to smaller food companies through safety and quality programs. We were very pleased to learn that the state was going to provide an annual grant for this consortium. In the summer of 2000, a $1.0 million grant from the State of Illinois was used to establish the Illinois Center for Food Safety and Technology (ICFST). Sizer appointed Matthew Botos as the director of ICFST. Botos had previously worked with Sizer at Tetra Pak

On June 9, 2000, Dr. Henney recognized me with the FDA's Special Citation; I am grateful to David Armstrong for nominating me for this honor.

I must now pause and describe the impact of Charles Sizer on the growth and mission at the NCFST. He had served as the center's director from 1998 to 2003, and his influence has continued to this day. Under his leadership, the food processing and packaging research programs were greatly enhanced. Our researchers were on the cutting edge of technology, focusing on topics such as high-pressure processing (which, instead of days, can now keep foods like store-bought guacamole fresh for weeks) and magnetic thermometers to permit real-time tracking of the temperature inside foods, such as Campbell's Chunky Soups, to prevent food-borne pathogen outbreaks. Other research areas included the aseptic processing of particulates, polymers in the prepackaging of food treated with radiation, the ultraviolet processing of juices, pulsed electric field processing technology, and several other novel sanitation methods. The Packaging Task Force and the Alfalfa Sprout Task Force were established under Sizer's watch at the NCFST. In addition, a new graduate program in food processing technology was introduced at IIT.

In the spring of 2000, Illinois Senator Dick Durbin visited the NCFST; he was greatly impressed by what he saw at the center. Ever since his visit, he has been a great supporter of its mission; he even noted, "What your scientists' experience in processing and HACCP [Hazard Analysis and Critical Control Point] gives them the tools to find and evaluate promising technologies." Everyone at the center appreciated his kind words and support, as he was helping us all move forward on our mission of creating a safer, healthier food supply.

In 2003, Karen Carson stepped down as the project officer, and Don Zink succeeded her. Once again, I managed the collaborative center until the new director, Martin Cole, joined us in 2004. Prior to assuming this position, Cole was chief scientist and deputy chief executive officer of Food Science Australia. He had a long history of holding senior positions within the industry, including serving as the group director of food safety for Nabisco, one of the center's member companies in the U.S. He already was familiar with the collaborative activity at our center.

September 11

September 11, 2001, had a most powerful impact on the entire nation. The terrorist attacks on the World Trade Center and the Pentagon were the beginning of a sad new era in the US. These attacks, combined with other threats such as the letters containing anthrax and Bovine Spongiform Encephalopathy (BSE), or Mad Cow Disease, heightened our concerns about the safety and security of the nation's food supply.

The members of the NCFST sprang into action. Cole and his executive leadership team added counterterrorism research to its agenda. NCFST researchers delved into the potential hazards and behaviors of nontraditional contaminants and emerging pathogens in food matrices. The FDA granted us additional funds to construct a Biosafety Level 3 Food Processing Biocontaminant (BSL-3) pilot plant and a microbiological laboratory. I was pleased to contribute

$1.0 million toward this $6 million facility. My contribution came from the rental money that the FDA had paid us for the use of the research facility, which I had saved over the years as the center reported to me.

The BSL-3 facility consisted of a laboratory and a pilot plant in which collaborative and contract research and validation of processes could be conducted, by simulating real-world applications with large-scale equipment and large volumes of organisms. On November 30, 2010, this facility received certification, acknowledging that it was meeting the guidelines of the CDC/NIH Biosafety in Microbiological and Biomedical Laboratories. The NCFST had officially joined the fight against global terrorism.

Director Martin Cole

The NCFST flourished under Martin Cole's leadership. He established the Executive Advisory Board (EAB), which replaced the OAC. This committee, on which I served, included some of the industry leaders from the center's member companies; it currently advises the center's director. The NCFST developed five science platforms on which its food safety and security research were to be conducted: chemical contaminants and allergens, food defense, food microbiology, food packaging, and food processing. High-pressure processing research gained momentum at the NCFST, and the center's consortium won the Institute of Food Technologists' (IFT) Innovation Award, for its pressure-assisted thermal sterilization process. Under Martin's leadership, a new nutrition research program was also initiated at the center. Britt Burton-Freeman, known for her studies on human nutrition and its impact on the body, was appointed director of the Center for Nutrition Research.

David Armstrong, the FDA associate research director, retired in 2005. Catherine Nnoka then joined the center as the new director of operations support. Cole left the NCFST at the end of 2009 and returned to Australia. Before he left the center, Cole bore witness to the next stage in the NCFST evolution as a leader in food safety research.

In 2009, the EAB recognized the success of the center over the past 20 years and recommended elevating the center to an "institute." This was, no doubt, partly due to the extramural grants at the center that amounted to over $12 million (see figure 5-2 that highlights the successful collaborative work carried out at the NCFST). The EAB suggested renaming the collaborative research facility to the Institute for Food Safety and Health (IFSH), believing that this would greatly expand the NCFST's capabilities.

After Cole left, I once again found myself serving as the interim director. I appointed Jason Wan as the interim associate director. Wan, who had worked previously with Cole in Australia, had joined the center in 2009. He, Rich McDonald, and I managed the center for about 10 months. We submitted to the FDA a five-year renewal proposal for the NCFST. Meanwhile, I chaired the search committee for the new director. The search committee was comprised of three members of the EAB, three members from the FDA, and three members from IIT. The committee identified Robert E. "Bob" Brackett, a former FDA/CFSAN director, as the top candidate for the director's position.

Don Zink, at the FDA, and I both met privately in Washington, D.C., with Bob Brackett, who at the time was serving as senior vice president and chief science and regulatory officer at the Grocery Manufacturing Association (GMA). Prior to joining the federal government, he had worked at the Univ. of Georgia.

At this time, I proposed to IIT President John L. Anderson, that the NCFST be elevated to an institute. I also pressed for Brackett to be given the title of vice president, in addition to being named the institute's director. Anderson agreed to my request, but he wanted the new institute to report to Provost Alan W. Cramb, just like the heads of the other institutes at IIT.

Brackett was pleased to accept the offer of vice president and director. Shortly after he joined IIT, he applied for the rank of full tenured professor. It was granted after the university's committee on tenure and promotion reviewed his complete professional dossier, which included several letters of support from food science leaders in both academia and industry. I am happy to say that Brackett is the first director who, in addition to being the director, is both a university vice president and full professor.

Institute for Food Safety and Health (IFSH)

When Bob Brackett joined us at the beginning of the fall semester of 2010, he immediately went to work to plan the transition of the NCFST to the IFSH. On December 8, 2010, the IIT Executive Committee of the Board of Trustees approved the institute's new status and its name change to the IFSH. The institute was dedicated in April of 2011. When Brackett began, he needed a strong associate director. McDonald and I pointed out Jason Wan's excellent qualifications, whom Brackett promptly promoted to be the new associate director of IFSH. Armand Paradis was recruited by Cole to serve as the director of business development at IFSH.

Brackett has since restructured the organization to meet the aims of all its members and stakeholders in industry, government, and academia. In addition to the FDA's Division of Food Processing Science and Technology, he also reorganized IFSH into four centers: the NCFST, the Center for Processing Innovation, the Center for Nutrition Research, and the Center for Specialty Programs. Its two graduate education programs are now housed in the newly formed Department of Food Science and Nutrition in the university's School of Applied Technology, where C. Robert "Bob" Carlson currently serves as the school's dean. Today, approximately 100 graduate students are enrolled in courses and programs in the Department of Food Science and Nutrition. In 2017, a new Ph.D. program in food science and nutrition was added.

Today, the annual meetings of the IFSH are heavily attended by industry, and the attendees hear presentations from FDA leaders, IFSH scientists, and industry representatives. Currently, over 60 industry members are involved with the IFSH, and some 100 staff members work there, including about 50 staff members in the FDA's Division of Food Processing Science and Technology, still housed at the IFSH.

More recently, the FDA awarded a $5.8 million special grant payable over three years (2015–18) to IIT to establish the Food Safety Preventive Controls Alliance (FSPCA), a public/private alliance consisting of key industry, academic, and government stakeholders. The main mission of FSPCA is to develop standardized core curricula, training, and outreach programs, to assist both domestic and international companies in complying with the preventive controls regulations that are part of FDA's Food Safety Modernization Act (FSMA).

Under Brackett's management, IFSH has continued to build on the successful foundation of the NCFST. By tapping into the NCFST researchers' expertise, the FDA has benefited by developing its regulations for food safety, and industry stakeholders have proactively resolved important path-to-market food safety and quality issues. IIT researchers have been able to utilize their talents to the fullest, making the nation's food supply safer, and have sharpened their skills through investigation of novel programs.

I am also pleased that the IFSH has established the Darsh Wasan Award in Food Safety to recognize the achievements of individuals who have made outstanding contributions to food safety across government, academia, and industry.

IIT has received nearly $150 million in grants from FDA in support of the NCFST/IFSH since its inception in 1988.

The Moffett Center with Buildings 90 and 91 in the foreground in the late 1980s

From the left: Isabel Wolf, Darsh Wasan, Dick Lechowich, Dick Snow, and David Armstrong in

front of the main entrance of the Moffett Technical Center in 1988. The sculptor of Rockefeller Center's *Atlas* created the striking bas-relief sculptures for Corn Products Company (later CPC International), the original owners.

At the NCFST dedication and ribbon-cutting ceremony in May 1991

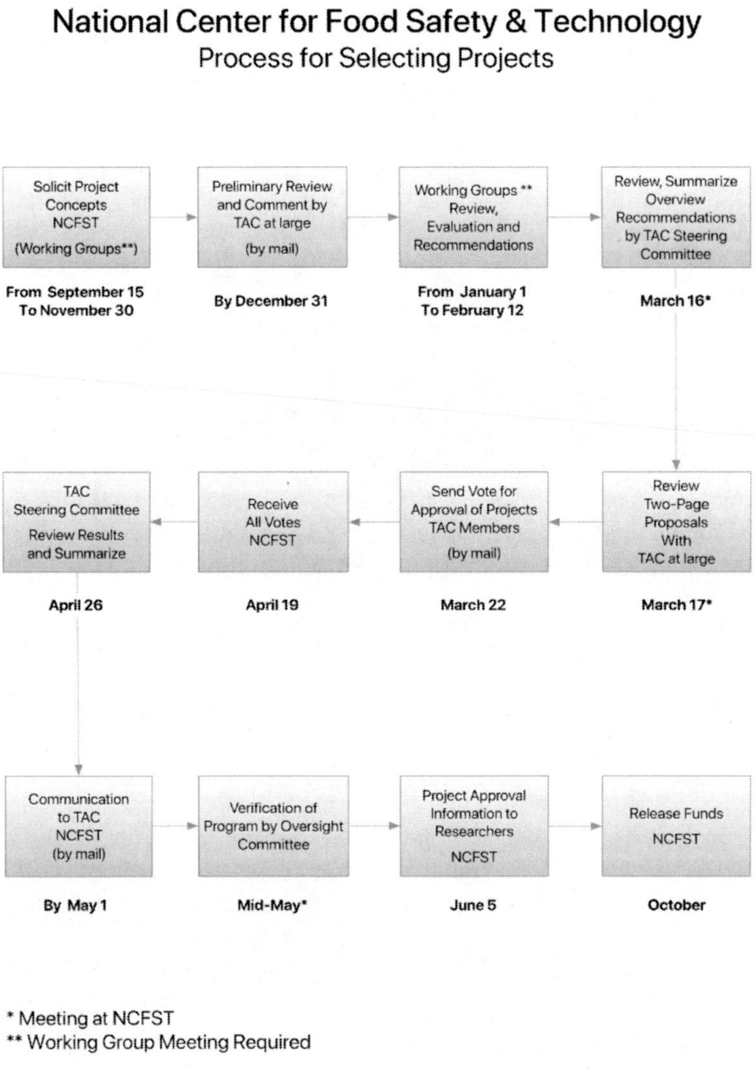
National Center for Food Safety & Technology
Process for Selecting Projects
Solicit Project Concepts NCFST (Working Groups**)
From September 15 To November 30
Preliminary Review and Comment by TAC at large (by mail)
By December 31
Working Groups ** Review, Evaluation and Recommendations
From January 1 To February 12
Review, Summarize Overview Recommendations by TAC Steering Committee
March 16*
Review Two-Page Proposals With TAC at large
March 17*
Send Vote for Approval of Projects TAC Members (by mail)
March 22
Receive All Votes NCFST
April 19
TAC Steering Committee Review Results and Summarize
April 26
Communication to TAC NCFST (by mail)
By May 1
Verification of Program by Oversight Committee
Mid-May*
Project Approval Information to Researchers NCFST
June 5
Release Funds NCFST
October
* Meeting at NCFST
** Working Group Meeting Required

From the left: David Armstrong, Jane Henney, Charles Sizer, and Darsh Wasan at the NCFST in 1999

Food and Drug Administration

Commissioner's Special Citation

PRESENTED TO

Darsh T. Wasan, Ph.D.

For extraordinary vision and creativity in developing the Nation's first food safety consortium joining government, industry, and academia- the National Center for Food Safety and Technology.

June 9, 2000

DATE

COMMISSIONER OF FOOD AND DRUGS

Special Citation from the FDA, signed by Dr. Henney

Commissioner Henney (far right) presents the FDA Commissioner's Special Citation to me on June 9, 2000.

From the left: IIT Vice President David Baker, Senator Dick Durbin, Darsh Wasan and Charles Sizer in 2000

From the left: Martin Cole, Bob Brackett, Senator Dick Durbin and Darsh Wasan at the NCFST in 2007

At the IFSH ribbon-cutting ceremony in April 2011

NCFST/IFSH Collaborative Works

1990–2017

- 1990 Regulation for safe packaging of fish
- 1991 Guidance for replacement of mercury thermometers with electronic sensors
- 1992 First 2-hour E. coli test for juice, milk and beef
- 1993 Guidance for control of mycotoxins in apple juice
- 1997 First aseptically packed complex food accepted by FDA
- 1998 Validation guidelines for automated process control
- 1999 Best practices guidance for allergen control and cleaning protocols
- 2000 Most comprehensive database on food botulism
- 2001 Juice HACCP guidance and training program
- 2002 Process to control acrylamide formation
- 2006 Guidance for cooked eggs during avian flu outbreak
- 2009 FDA acceptance for pressure assisted thermal sterilization (PATS)
- 2009 Application of Food Safety Objective concept to commercially sterilized products
- 2011 Inauguration of IFSH
- 2012 Low Moisture Food Safety Consortium
- 2013 Department of Food Science and Nutrition within the School of Applied Technology
- 2014 BSL-3 Pilot Plant Accreditation
- 2015 Food Safety Preventative Controls Alliance (FSPCA)
  Sprout Safety Alliance (SSA)
  FDA Acceptance of Pressure Enhanced Sterilization (PES) filing
- 2016 Nation-wide and International FSMA training programs
- 2017 Ph.D. Program in Food Science and Nutrition
  ISO 17043 Accreditation for Laboratory Proficiency Test Program

Chapter 6
The Office of Research and Technology (1988–91)

The summer of 1988 was a time of anticipation and change at IIT. I was about to complete my assignment as interim dean of engineering, but I had not yet received any news from the FDA about a grant to establish the NCFST at the Moffett Campus. It was in this period that President Meyer Feldberg began to implement his vision of IIT; he announced that he was creating a new position, the IIT/IITRI vice president of the Office of Research and Technology.

The functions of this new office were to promote and encourage research at IIT and to act as the institutional interface between IIT and the IITRI that would deliver benefits to both institutions, such as being able to better coordinate proposals and projects and boost marketing. With the university entering its second century, these joint efforts would take on a new significance in defining the future of IIT. The university would be able to continue its role as one of the nation's premier research and education centers well into the next century.

Feldberg appointed a four-member search committee that included Provost Mel Bernstein and Professor Hassan M. Nagib, both of IIT, and President David L. Morrison and Narayan P. Murarka, both of the IITRI. I was one of two candidates who interviewed for this position; the other was Linda Cooper of the IITRI. Effective September 1, 1988, I was selected to establish the new Office of Research and Technology as the vice president of the IIT/IITRI.

In hindsight, my selection for this position may largely have been due to my role in successfully securing the gift of the Moffett Technical Center. At almost $7 million, the center's gift from CPC had been a major windfall for the university. My selection may also have been partially aided by my position as the principal investigator in the successful submission of the $3.7 million grant proposal to the FDA to establish, in partnership with the IITRI, the NCFST.

The core of Feldberg's initiative was to inspire active collaboration between the basic and applied researchers at IIT and those at the IITRI. I was given two offices—one office next door to the office of IITRI President Morrison on the top floor of the IIT Tower and a second office on the second floor of Perlstein Hall (formerly occupied by IIT President Thomas L. Martin Jr.). Today, I still occupy the office in Perlstein Hall; fortunately, my research lab is also located on the first floor of this building. One-half of my salary was paid by the IITRI and the other half by IIT. I was formally reporting to IITRI President Morrison and, whenever appropriate,

IIT Provost Bernstein while also serving as a member of IIT President Feldberg's council.

Prior to the establishment of this new office, the Office of Sponsored Research at IIT was reporting to the dean of the graduate college. To bolster the research enterprise, the first task I undertook was to integrate the various strategic functions to enable efficiency and to encourage and support the creative efforts of IIT faculty members and graduate students. This approach included a close examination of several offices: the Office of Research Administration, the Technology Commercialization Center, the Research Council, the Patent Council, the Institutional Review Board for Human Subjects, the Joint IIT/IITRI Research Collaboration, the Center for Environmental Research (CER), and the newly established National Center for Food Safety and Technology (NCFST). All of these disparate offices were eventually to be organized and streamlined into one coordinated body (see figure 6-1 for the Office of Research and Technology organizational chart).

I recruited Thomas M. Jacobius, from the IITRI, to head the Technology Commercialization Center, and Irv Fieldhouse, also from the IITRI, to both serve as senior advisors to stimulate the Joint IIT/IITRI Research Collaboration.

The IIT research and technology enterprise represented a highly leveraged approach with a very limited staff investment; a few key staff positions were to guide all of its activities. This arrangement assured a cohesive-team approach that was mutually supportive, with a flat organizational structure and straightforward channels of communication. Such an agile team approach also assured that faculty needs and research possibilities surfacing in one area would be supported across the entire research enterprise to ensure that maximum impact and collaborative value were realized.

From 1988 to 1991, the Office of Research and Technology's level of funding for grants and contracts rose from $8 million to $12 million; during this same period, there were only six core staff members associated with the enterprise. The Office of Research Administration and the Technology Commercialization Center were merged into one and managed by Jacobius, with only one professional support staff member and one administrative assistant. There were no reductions in the quality or the level of the services provided. Also, there was a 24/7 commitment to fully supporting the needs of the faculty and the goals of the university. Due to the ongoing budgetary challenges, a continued effort was also made to reduce expenses while maintaining the quality of the services by continually evaluating the costs and the benefits of various budget elements. Implementing a culture of fiscal responsibility and a more entrepreneurial mindset helped us sustain, and actually grow, the impact that the IIT research enterprise was able to achieve.

Throughout the first year, I visited the top management at various locations of the IITRI— in particular, Annapolis and Lanham in Maryland and the National Institute of Petroleum and Energy Research in Bartlesville, Oklahoma. These visits proved very helpful in identifying opportunities for joint IIT/IITRI research and educational projects. The meetings that we held also served to bring the staff at off-campus IITRI locations into the ongoing collaborative process. At the time, the IIITRI had over 1,000 employees.

Several joint meetings were held, and the top management of the IIT/IITRI Center sought to continuously define and develop strategic technical areas of business opportunities. Five technical areas were targeted for collaboration: energy, environment, food safety and technology, manufacturing and materials processing, and transportation engineering (including railroad engineering). In addition, the Education and Research Initiative Fund (ERIF) was established with financial support from the IITRI.

Education and Research Initiative Fund

The Education and Research Initiative Fund (ERIF) was established to encourage development of innovative, high-quality programs in undergraduate and graduate education and various associated research. The intent was to provide seed funding to individual faculty members or to groups of faculty who wanted to carry out exploratory programs related to, for example, curriculum development, disciplinary and interdisciplinary research, or the identification of other unique quality opportunities on campus. In March 1990, 16 new projects were awarded funding totaling nearly $173,000.

Technology Commercialization

The Technology Commercialization Center was funded through a grant from the State of Illinois. It provided the seed money to technology-oriented projects that had reasonable short-term commercial potential. This center funded 13 projects during 1989–90.

In November 1989, the IIT/IITRI Office of Intellectual Property Management was established to encourage invention disclosure by faculty and staff, and it offered recommendations about patentability and licensing potential. This office also promoted IIT- and IITRI-patented technologies to potential licensees. During the first year, three IIT faculty members submitted nine invention disclosures to the patent office, with one leading to a patent application. Selected IIT/IITRI patents were also presented at a meeting of corporate and university researchers sponsored by Technology Transfer Conferences, Inc. and the Purdue Research Foundation. In cooperation with Technology Marketing Associates, Inc., a powerful new concept was developed for the bold and sustained marketing of the IIT/IITRI Center research. Called IIT FOCUS, this new concept incorporated an interdisciplinary sponsored-research program, involving world-class experts into an interactive focus group. The program brought together leading experts in research areas such as energy, environment, food safety and technology manufacturing, and transportation from industry, government, and academia.

Joint IIT/IITRI Research Collaboration

One of the primary purposes of the Office of Research and Technology was to encourage and facilitate joint research ventures between IIT and the IITRI. The synergy would benefit both institutions, as well as their clients and the markets they sought to address. Irv Fieldhouse, senior advisor at the Joint IIT/IITRI Research Collaboration, worked closely with the Office of Research and Technology, providing invaluable services in identifying opportunities for IIT/IITRI joint collaboration, convening meetings to discuss these opportunities, and assisting with proposal preparation. Fieldhouse was an expert at handling all such matters involving increased support for IIT's research and educational programs. (See figure 6-2 for highlights of the IIT/IITRI collaboration during the fiscal year 1989–90.)

Transition Period

In 1989, President Feldberg left IIT to become dean of the business school at Columbia University; Henry R. Linden was appointed interim president. Initially, Linden was somewhat reluctant to accept the terms proposed to him by Bob Pritzker, chair of the IIT Board of Trustees. I convinced Linden to accept those terms, but only for the interim appointment. At that time, I revealed to him that, without his knowledge, I had originally nominated him for the IIT presidency in 1974, when Thomas L. Martin Jr. was offered the position.

Following the appointment of Linden as interim president, the Board of Trustees decided to conduct a national search for a qualified candidate. A search committee was formed, and I was asked to serve on it. Mel Bernstein, IIT provost, and Lew Collens, dean of the Chicago-Kent College of Law, both applied for the position. In addition to interviewing them, the search committee also interviewed several external candidates. For a variety of reasons, the committee finally rejected all of the external candidates and recommended the two internal candidates, with Lew Collens as the next president and Bernstein as the chancellor of the university.

Collens had obtained his bachelor's and master's degrees at University of Illinois at Urbana-Champaign and had a J.D. from the

University of Chicago. He began serving on the IIT faculty in 1970. Under his guidance, the Chicago-Kent College of Law launched novel programs to prepare students challenged with modern, complex legal issues.

In 1991, Bernstein left IIT to join Tufts University; President Collens then formed another search committee to recommend an internal candidate as the next provost. The committee interviewed three candidates: Joel Goldhar, former dean of IIT Stuart School of Business; Hassan Nagib of the mechanical and aerospace engineering department; and me. After careful consideration of the qualifications of the nominees, President Collens appointed me as the next provost. I began serving as provost in May 1991.

Office of Research and Technology Organization Chart

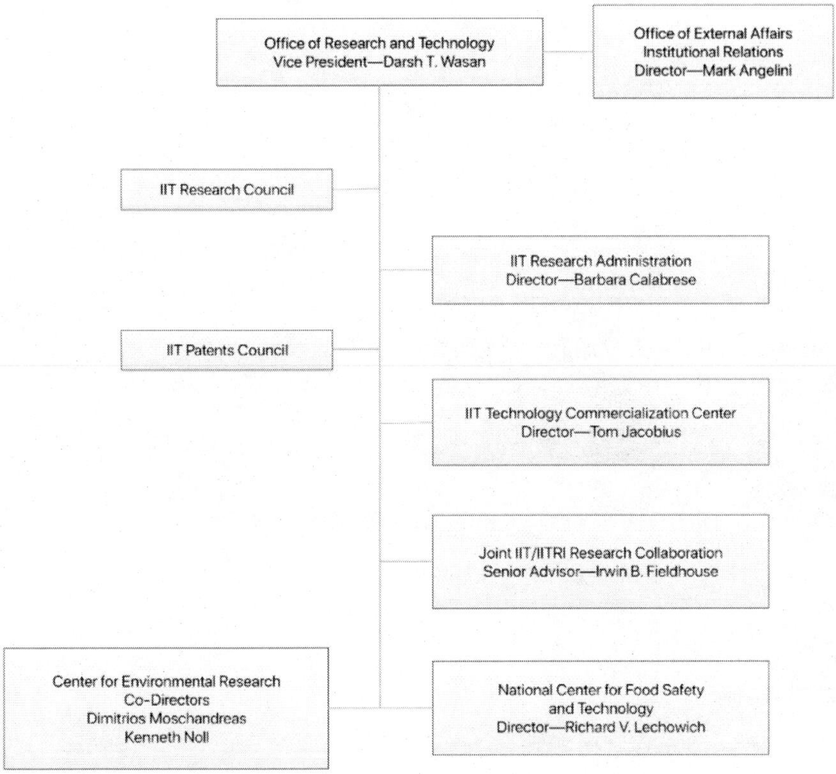

With President Lew Collens in 1991

Highlights of the IIT/IITRI Collaboration FY 1989–90

- Nine contracts were awarded totaling $28.96 million; five of these contracts resulted from the proposals written in FY89. Total funding included a $24 million contract for the Instrumented Factory for Gears (INFAC).
- Twenty-seven joint proposals were funded totaling $25.8 million.
- IIT/IITRI faculty and professional staff were involved in 20 joint projects.
- Seven operation synergy grants were awarded totaling nearly $80,000 in institutional funds.
- An IITRI Graduate Fellowship Program was established at IIT, which included four $25,000 per-annum fellowships.
- A workshop was held, at which some 50 IIT/IITRI staff members focused on the objectives, direction, and future goals of the IIT/IITRI Center for Environmental Research (CER).
- Five technical areas of opportunity were targeted and developed for joint collaboration between IIT and the IITRI: energy, environment, food safety and technology, manufacturing and material processing, and transportation engineering.
- A potential extension of the IIT curriculum in electrical engineering and computer science was identified by establishing a branch of the IITRI/ECAC in Annapolis, Maryland, with an estimated enrollment of 100 students per semester.
- A promotional effort regarding the use of hydrogen as an energy-storage and transport medium and transportation fuel was initiated.
- The possibility of forming of a national academy of railroad engineering and management at the IIT/IITRI Center was considered.

Chapter 7
Office of the Provost (1991–96)

Whenever they think of the early 1990s, many Chicagoans will fondly recall how, with their first three NBA championship titles, the world-class Chicago Bulls energized the city. It was their first of two three-peats; aspiring athletes everywhere wanted to "Be like Mike!"

Iraq's invasion of Kuwait in August 1990 caused the Gulf War to become the backdrop of our political life; this was soon overshadowed by the collapse of the Soviet Union. In 1992, Bill Clinton defeated George H. W. Bush and H. Ross Perot, becoming the 42nd President of the United States. The nation was facing increasing acts of terrorism on its home soil, notably the attack on the World Trade Center in 1993 and the Oklahoma City bombing in 1995.

The North American Fair-Trade Act (NAFDA) crafted a new view and approach to the globalization of business relationships. The Hubble deep-space telescope was launched, revealing the beauty of our mysterious, immense universe. With his paper on carbon nanotubes, physicist Sumio Iijima drew the attention of scientists to carbon nanotubes (buckytubes) with novel applications in energy storage and electronic devices. The computer scientist and engineer Tim Berners-Lee invented the World Wide Web.

The world was awash in constant change, heavily impacting our views and funding of education; it was an exciting and challenging time to be in academia. IIT itself was in the middle of a significant development in how education was being used to serve society. In January 1991, the Daniel F. and Ada L. Rice Campus opened its doors in Wheaton.

The Provost: Captain of Change

The Office of the Provost is vital to the success of a university; it is the operational captain managing all academic affairs and responsible for strategic planning. When I became provost in 1991, I soon realized that in terms of its future and its focus on research, IIT was at a crossroads. The educational environment in which the university long existed was becoming increasingly selective in its student body and its research mission.

The interaction between applied research and the basic sciences was increasingly becoming a more important factor in maintaining IIT's competitive position. I also thought that its competitive edge could be sharpened by reducing some obvious inefficiencies and by strengthening the school's focus on its students and on research.

A part of my approach to these challenges was to combine some departments. My plan to streamline the school may have been seen by some as revolutionary, but I believed it was a necessary step. I merged the departments of biology, chemistry, physics, computer science, and mathematics into the new Armour College of Engineering and Science; these disciplines had all previously been part of Lewis College. This move paid off quickly, because there were some immediate savings in administrative costs. Once the hard sciences were consolidated into one administrative unit, I nudged the departments of humanities, social sciences, and psychology to work together under their own separate unit, which eventually became a part of the renamed Lewis College.

IIT for the Students

Although it was useful and efficient to streamline various departments into two separate divisions, this was only the first step toward creating a world-class academic institution. Moreover, the leadership had to strongly believe in the future of the institution, and I knew well that we needed great visionaries capable of leading IIT in the right direction.

I made several significant appointments during my time as provost. Geoffrey Higgins agreed to become vice provost; he was well known for his work in materials science and in 1979 had taken home the Literary Award of the American Nuclear Society. Stephen M. Copley of the metallurgical and materials engineering department was tasked to serve as the dean of the Armour College of Engineering. Tom Jacobius became the director of research administration, and Sohail El Baz assumed the directorship of the Paul V. Galvin Library.

Our students were becoming increasingly diversified, thus making it imperative for us to recruit a faculty that truly reflected their needs and to respond to the changing workplace. To serve as role models, IIT President Lew Collens and I hired 11 women and minority faculty members.

Just as they need to figure out their budgets, institutions must also continually grapple with the best use of money and resources. With this approach in mind, Copley helped me in the development and presentation of an economic model of IIT. Delivered to the upper levels ofthe administration, the model would enable the administration in determining the real cost of our academic programs, as well as correctly evaluating the income generated from eachprogram.

While the reorganization of the departments and the process of hiring new faculty were keeping me busy, there was still much to do. Being the provost was, of course, important, but I also was still committed to personally helping students succeed in their academic disciplines. I continued mentoring eight to ten graduate students of my own, and I also served as vice president for research and technology; thus I knew firsthand about cutting-edge research as well as the varied demands of the business community.

I strongly believed that students must have the opportunity to work on real-world problems and to actively cooperate with industry. I helped the architecture faculty secure a

$240,000 grant from the Korean industrial giant Hyundai for an initial one-year collaborative research project: to develop a conceptual design for what would have become the Hankang City skyscraper in Seoul, the tallest building in the world. At nearly 120 stories high, it was to house stores, businesses, and a hotel. A team of three graduate students and professors, led by Mahjoub M. Elnmeiri, worked with researchers from Hyundai to work on this multifaceted project. Although the building never graced the Seoul skyline, the experience gained by the students and staff was invaluable.

There was a strong and constant need to provide our students with innovative opportunities to explore and learn. The athletic department recruited, scheduled, and developed three new varsity sports teams: women's varsity swimming, women's junior varsity volleyball, and women's varsity cross country. These teams helped us recruit students, adding another dimension to student-life activities. Athletics increased the total number of our student-athletes by 48 percent over the previous year.

Encouraging more female athletes to attend the university helped create a well-rounded educational environment at IIT, with its large percentage of international students. It was vital to focus on how best our students from different backgrounds would understand each other, and in ways that were instructive and healthy. We developed and implemented 18 university-wide workshops on IIT's cultural awareness, focusing on such issues as race relations, racism in higher education, sexual harassment, gender equality, affirmative action and reverse discrimination, Asian relations, cultural diversity, and the language of free speech and hate.

Giving Teachers a Helping Hand

Although high-quality instruction, with its various teaching models and methods, is always essential, students still require knowledgeable guiding lights, indicating the need for investment in our faculty. This awareness led to the creation of the first workshop on teaching effectiveness; held at the Rice Campus, the workshop attracted over 65 faculty members. I developed a faculty career and reward structure, enhancing productivity and encouraging and rewarding quality teaching and instructional scholarship.

Leon Lederman, a Nobel Laureate physicist, joined our faculty as the Pritzker Professor of Science. We asked him to teach a freshman physics course. He was very helpful in establishing the campus-wide Board of Trustees' Annual Lectureship Series. He was also instrumental in attracting prominent speakers to IIT, such as Arthur E. Goldman, a NASA administrator who worked in the Space Shuttle program.

During my time as provost from 1991 to 1996, the university created the title of Distinguished Professor, an honor to be awarded solely for exceptional achievement by full professors. As at other universities, the candidates had to have attained national or international preeminence in their fields based on scholarly work and excellence in teaching. The first two selected were professors Sheldon H. Nahmod, who later founded the Institute for Law and the Humanities, and A. Dan Tarlock, who later became a special legal advisor for the NAFTA Commission on Environmental Cooperation.

Establishing IMCA-CAT and Bio-CAT

Helping our faculty obtain grants and performing the type of work that would lay the foundation for progress in both science and business were integral to their professional growth and an important way to serve society at large. However, it can be challenging to find a person who will be the best candidate, so having an advocate for that candidate can make a significant difference.

In 1992, I vividly remember meeting and interacting with 12 managers and research directors of drug companies on a Saturday morning at O'Hare airport in Chicago to form a consortium at IIT. The Industrial Macromolecular Crystallography Association-Collaborative Access Team (IMCA-CAT) had an initial funding of about $7 million to build beam lines at Argonne National Lab's advanced photon source in order to identify the crystallographic structures of potent drugs. Timothy I. Morrison, a physics professor, became the first director of this project; he was a former U.S. Air Force service member and had come to IIT from Argonne.

In 1995, I also assisted Grant B. Bunker, a junior physics professor at the time, in receiving the first five-year grant from the NIH for $8.6 million to establish the Biophysics Collaborative Access Team (Bio-CAT) at IIT. Although Bunker was the perfect leader for this work, the proposal review panel had some reservations due to his inexperience in managing a large grant. I flew to Washington, D.C., and convinced the panel that Bunker would excel as the principal investigator of this grant. We received the grant, and he indeed did a superb job.

Developing the School and Its Mission

One of the primary aims of IIT has been to serve its students, teachers, and society in general, ensuring that the facility and the programs offered were the very best possible. During my tenure as provost, my office established a number of programs that were responsive to market needs and would build on the strengths of our existing programs. We added a B.S. degree in computer engineering and renamed the department the electrical and computer engineering department. The enhancement of our computer infrastructure was a key part of this initiative. Other programs also were enhanced as IIT began granting B.S. and M.S. degrees in architectural engineering. We renamed the department the civil and architectural engineering department.

Because partnerships in learning had become essential, we offered B.S. and M.S. degrees in a joint engineering/M.D. program with Rush Medical College. Our business school opened its doors to a more diverse range of degrees by offering the world's first M.S. degree in financial markets and trading—an initiative suggested by then new Trustee John Adams "Jack" Wing, chief executive officer of The Chicago Corporation—and an M.S. degree in environmental management that rapidly gained national prominence.

We also began offering an M.S. degree in food safety and technology jointly through the departments of chemical engineering and biology, as well as an M.S. degree in technical communications and design through the Lewis College of Liberal Arts. In January 1992, a full- service M.B.A. degree program was initiated at the new downtown campus when the Chicago- Kent College of Law moved into its current home at 565 West Adams Street.

Quality Journey

IIT's mission of education is directly linked to the ability of students and their desire to serve others. What we teach in our classrooms and how we carry out our research activities must be conducted with the highest ethical standards. Consequently, our curricular initiatives continue to be centered on educating our students on quality, creativity, ethics, and leadership—or QCEL, an initiative that began when I was provost in the early 1990s. In a world that continues to be full of rapid change and globalization, these four key precepts to success continue to be as relevant today as they were back then.

Robert W. Galvin, chairman of Motorola's Executive Committee and the past chairman of IIT's Board of Trustees, was not only a highly respected giant in the world of business, but also the innovator behind the concept of Six Sigma, which was a unique approach to bringing quality products to the marketplace. Late in 1991, soon after the publication of his book, *The Idea of Ideas,* Galvin invited Lew Collens, M. Zia Hassan, and me to discuss how the university could encourage a greater appreciation and development of QCEL skills in our students. The meeting had a profound impact on our thinking; soon after, Collens named QCEL as precepts of the dynamics that all IIT students, faculty, and staff needed to carry out in their professional and personal lives.

The 1992 academic year began with a daylong faculty retreat on a Saturday morning at Motorola University, where the IIT faculty participated in discussions on QCEL topics with experts from Motorola. Following the retreat, the provost's office organized regular lectures, seminars, and discussion groups featuring speakers from IIT, as well as external sources. At mid-year, Joshua Hammond, president of the American Quality Foundation, presented an on-campus lecture on quality to faculty, staff, and students. Bernard F. Sergesketter, an IIT trustee, spokeabout personal quality. We also regularly held faculty luncheon discussion meetings, during which faculty members exchanged their QCEL teaching ideas. Each college of IIT held a retreat on these topics to assess the progress made and plans for the following year; the first university- wide plenary session was held in April 1993.

QCEL was steadily institutionalized through the Center for the Study of Ethics in the Professions (CSEP). Also, following a seminar on teaching effectiveness sponsored by the provost's office and held earlier in the year, the Center for Innovative Learning and Education was created with a generous gift from Virginia Rogers. The Rogers Foundation awarded seven grants to our faculty for projects to improve the educational experience of our students. Not surprisingly, all of this activity culminated in the creation of a specialized minor in QCEL at IIT.

The IIT/Nalco Partnership

One of the key milestones in the quality initiative took place in the fall of 1992. Lew Collens and Zia Hassan attended a symposium in Cincinnati on the TQM University Challenge, hosted by challenge sponsor Procter & Gamble Co. TQM (total quality management) is a customer-focused management approach comprised of eight pillars: a customer focus, total employee involvement, process-centered thinking, an integrated system, a strategic and systematic approach to the mission, continual improvement, data-driven decision making, and communication.

After returning from that event, Collens contacted Steve Weber, president of Nalco Company's Waste Water and Treatment Division, to discuss how IIT and Nalco could form a business/university partnership to train our students and faculty in TQM methodology. An exploratory meeting, which Collens and I attended, led to IIT's successful proposal three months later for the IIT/Nalco Partnership for the TQM University Challenge.

This was a nationwide effort in which industry linked up with universities to transmit the lessons of quality management learned by industry over the last decade. By becoming a part of this effort to form a business and educational partnership, IIT joined a nationally recognized selective group of companies and universities who had made a significant commitment to improving the performance and quality of American higher education.

IIT's successful partnership with Nalco began in March 1993 with the establishment of the Quality Steering Committee for IIT's quality journey. The partnership was launched with the drafting of IIT's mission statement on quality and the identification of the first quality project. This project focused on improving the satisfaction of students with their financial aid, student billing, and registration processes. Training the staff in these areas led to the formation of three quality teams and a task force to coordinate their efforts. The task force produced significant changes over a short period of time; the sustained insistence of Collens on quality had underpinned much of the improvement in operations that took place at IIT.

IIT's quality journey initiative was comprised of four major themes: the quality, creativity, ethics, and leadership (QCEL) initiative; the creation of the Center for Innovative Learning and Education; the quality of operations (the IIT/Nalco TQM partnership); and the quality of IIT's overall mission and its future objectives. As a center of higher learning, the university always had one eye on the present, solving pressing problems and educating students, and another eye on the future.

Around this time, IIT itself had an urgent problem that arose during the 1980s; student enrollment was in decline, presenting a serious challenge to the university's future. However, IIT had always been home to some very innovative thinkers, and just when it seemed that the university might need to face serious changes to its mission, the future cast a ray of hope on the present. Part of this future included the National Commission for IIT.

IIT National Commission

In May 1993, President Collens and the Board of Trustees established the National Commission for IIT. Its charge was to understand what we knew, what we did best, what we should do to add value to IIT's strengths, and how to get better organized in order to maximize the university's financial viability in the future.

The Board of Trustees established a quick-paced agenda, requiring the commission to complete its deliberations within one year. Chaired by Robert Galvin, the commission had 20 members drawn from industry and academia and included IIT trustees and IIT faculty members. As provost, I participated in the commission's deliberations, where our discussions centered on four areas: academic distinction, student experience, campus and community development, and financial growth. Both Pritzker and Galvin provided gifts of $60 million each. These very generous donations would go down as a turning point in IIT's history, serving as lead gifts for a successful $250 million philanthropic campaign that created financial stability for the university and enabled the university to establish the Camras Scholars program and build the State Street Village student dormitory and the McCormick Tribune Campus Center. These major additions helped attract a talented student body and made the university a physically more attractive place for students.

Although the new funds helped create a stronger school overall, I believe that the most significant recommendation that the commission made was the development of a distinctive interprofessional educational initiative to serve the needs of undergraduates, enabling them to become successful 21st-century professionals. This initiative emphasized the importance of experiential learning, interdisciplinary work, and team skills—and ultimately resulted in the creation of the Interprofessional Projects (IPRO) Program (discussed in more detail below).

Enhancing the Computing Infrastructure at IIT

The Internet had profound implications for education and innovation, and I knew well that the future of our students was closely tied to their use and understanding of computer technologies. With the assistance of U.S. Congressman Dan Rostenkowski of Chicago, a grant of

$5 million was earmarked for IIT's use in the Department of Defense Appropriations Act of 1993. We were, however, later informed by the Office of the Director of Defense Research and Engineering that the responsibility for determining a suitable project to be funded with this amount was actually assigned to the Office of Naval Research (ONR).

A bit miffed, I visited Andre Van Tilberg, director of the computer science division at the ONR, curious to find out whether upgrading the computing infrastructure at IIT would be viewed as an acceptable use of these funds. He told me that this was indeed an appropriate project, but that our proposal first had to go through a technical evaluation by the ONR before the Office of the Secretary of Defense could grant the award and transfer funds to IIT.

This situation meant that we had to submit a proposal detailing our plans and include budget expenditures. After my visit to the ONR, I asked Professor Rollin Dix, the associate dean for computing at IIT, to prepare a draft proposal outlining our plans to upgrade the computer infrastructure. The next month, I took his completed draft proposal back to the ONR for review, noting also the comment made by Tilberg regarding the appropriateness of our proposal for funding by the ONR. In September 1993, we submitted the final proposal; I was soon very pleased to learn that we had received the entire grant.

Over the next three years, the DOD grant funds were used to enhance our computing facilities and the infrastructure on the main campus, now called Mies Campus. Funding would also specifically enable us to provide our students and faculty with state-of-the-art computer and network facilities for use in both instructional and research programs. Approximately one half of the funds were used to purchase high-performance computing and telecommunications equipment. The remainder was used to hire staff to install, manage, and maintain the equipment, as well as educate users, and to acquire and develop software and supplies for researchers, students, and participants in the IIT outreach program.

In addition, we modernized the IIT/V system, a high-speed delivery system for distance learning that enabled IIT to reach a larger student constituency. About a quarter of these expenditures directly supported IIT's educational programs through installation of computer classrooms, computerized lecture halls, a multimedia center, and a variety of educational software.

About 60 percent of the funds went to research groups for the purchase of multiuse workstation clusters, internal and external network interconnections, a research visualization center, an integrated library support system, and the cooperative use of Argonne National Lab's massively parallel computing system. The remaining funds were earmarked for the school's outreach efforts and for updating the IIT/V system with a high-speed network.

Interactive Instructional Network (IIT/V)

While distance learning is now commonplace, it was not always easy to deliver quality education outside the walls of lecture halls. Ready access to initiatives, such as massive online open courses (MOOC), was still far into the future. IIT/V first began broadcasting in the 1970s, and TV was a key tool for distributing educational materials and lessons; also, it was easily accessible to most students.

President Thomas L. Martin Jr., who joined IIT in 1974, was not only a pioneer in recruiting women and minorities, he was also a strong proponent of distance education. Prior to his arrival at IIT, he already had established an interactive instructional television network at Southern Methodist University in the Dallas area; it was a foregone conclusion that he would create a similar initiative at IIT as well.

IIT/V had begun as a microwave-based distance education network, allowing students to visit remote sites closer to home, rather than traveling into the city to listen to the lectures. In 1976, the school started broadcasting live courses in the greater Chicago area, from its main campus to the top of the Sears Tower (now the Willis Tower), and then on to several suburban locations. The one-way video, which was transmitted via the William F. Finkl Interactive Instructional Television Network (IIT/V), was directed to a variety of receiver sites, such as institutions of higher learning; public access sites; corporate sites, like Motorola and Caterpillar; and several other sites. (Finkl, then a member of our Board of Trustees, was the developer of new types of steels and was known as the father of quality control in the steel industry.) The one- way video system, with a call-in phone, enabled our remote students to participate in class discussions; also, the proctored exams at the sites allowed students to complete courses at remote locations.

In 1971, the FCC created the Instructional Television Fixed Service (ITFS) as an educational service to provide classroom instr uction to remote audiences. It granted IIT the rights to use two ITFS channels: the E–channel, granted in 1976, and the G-channel, granted in 1982–83. While we were able to reach a larger audience, we also underwent a slowdown in enrollment for distance education classes, because corporations had stopped providing tuition remission to their employees. This decline not only meant that fewer people would be engaging in new ideas while continuing their education, but also that the channels were being partially unused, clearly a wasted resource.

Louise Hewitt, then assistant dean for educational technology, and I began working with a cable company to lease the channels at specified days and times when IIT was not using them. IIT was able to negotiate with Clearwire/Sprint to lease both of our channels for a number of years. While it was in effect, the leasing deal produced an income of nearly $2 million per channel a year; in January of 2018, IIT finally sold off one channel to Sprint for $65 million. Chief Technology Officer Ophir Trigalo facilitated this transaction.

Over the years, we also tested new technologies, such as two-way video and audio over standard telephone lines and utilizing the Internet for even wider distribution. Today, by utilizing Internet technology, our extensive curricula can reach anyone, anywhere, providing the highest quality educational experience for both faculty and students.

Lessons Learned from the E³ Program

In the mid-1990s, trends in engineering education and feedback from industry leaders (notably, employers of our college graduates) suggested that there was a need to cultivate graduates whose credentials were purposefully broadened and strengthened through holistic education. Specifically, there was an expressed need to emphasize the development of competencies through the undergraduate experience, so that IIT graduates could then have a greater impact within their organizations and in their professions. The educational goals were for graduates to communicate ideas more effectively; to become stronger team players, leaders, and practical problem solvers; and to become comfortable with the ambiguity of open-ended challenges and opportunities.

For graduates, these efforts would often involve making nontechnical and ethical decisions, and, at the outset of their careers, hitting the ground running. Over time, it became clear that these holistic attributes were highly valued by numerous professional bodies spanning engineering, science, architecture, law, design, business, and psychology.

As I reflected on previously taken IIT initiatives, I thought about an experimental education initiative that we attempted to implement in 1972. Back then, Paul Torda, a professor of mechanical engineering, had secured a five-year grant from the NSF, entitled Education and Experience in Engineering (the E^3 program). The aim was to develop and demonstrate a selective program model involving 10 to 20 percent of the undergraduate engineering population. The students would be organized in groups, devoting their four years to problem solving through design projects, acquiring knowledge of fundamentals, and studying what they needed to know. There would be no formal courses to take or other requirements to fulfill; it was to be a bona fide totally project-centered learning experience.

As department chairman at the time, I assigned Professor Bernie Swanson, of the chemical engineering department, to work with Torda on this initiative. Torda was known for his *Blue Flame*, a rocket-propelled car, designed by his teams of students and faculty members, which, in 1982, held the record for the fastest car traveling at 622.4 mph.

While he may have been a great engineer, Torda's real passion was education; his commitment to students made him ideal for the leadership of the E^3 program. Other departments also assigned some of their faculty members to participate in this endeavor. Regrettably, when the grant ended, the E^3 model was not sustained or institutionalized by IIT.

The IPRO Program

When Lew Collens asked me about instituting a new initiative of interprofessionalism into the undergraduate curriculum, I briefed him on our E^3 experience and advised him that we should recruit

Professor Hassan Nagib, who had worked with Torda in the E^3 program, to lead this initiative. Upon my advice, Collens asked Nagib to respond to a recommendation of IIT's National Commission and develop plans to offer an educational initiative for interprofessional. In 1994, Collens, Nagib, and I held several meetings with a select group of trustees to brainstorm about this concept.

However, in the spring of 1995, Collens announced that beginning in the fall he was appointing Nagib as dean of the Armour College and vice president of the main campus. He also appointed Henry H. Perritt Jr., dean of the law school, as the vice president for the downtown campus; my title was changed to provost and senior vice president.

Subsequently, I asked Tom Jacobius to assume a new role to support the development of the Interprofessional Projects (IPRO) program. He first developed a business plan for its operation and succeeded in also attracting funding from the NSF to explore the IPRO vision, its pedagogy, and the practicalities associated with institutionalizing the program. I appointed Professor Mohammad Shahidehpour, of the electrical engineering department, as the director of sponsored research (replacing Jacobius) and as dean of the graduate college.

Building on his experience on the previous initiative with Torda, Nagib successfully led the initial development of the IPRO course from 1995 to 1998. The first prototype team project, guided by Nagib and Professor Kevin P. Meade, was entitled: Technical, Economic, Social, and Legal Perspectives on the Barriers to Diffusion of High-Speed Rail Technology in the United States. Several faculty members from various departments also collaborated on this project.

The journey that led to the prototype IPRO course, and its continuous pilot testing, scaling-up, and institutionalization, had begun with the National Commission. Ultimately, it led to the establishment of IIT's 1994 strategic plan. Collens's desire to use IPRO for improving the academic experience for students has given back much to the university. The students and faculty of various disciplines began to fully collaborate, and the students learned much from each other. In addition to IPRO, the National Commission also had called for the university's international expansion. The efforts during the early to mid-1990s set the stage for IIT's progress into the approaching millennium.

The school was on the cusp of a radically new world, where education had to immediately address the needs of students, as well as of science, business, and society at large. While no one could be sanguine that they have done everything possible to prepare students for the future, I was confident that IIT's leadership had indeed positioned the school to be able to embrace the coming challenges and the myriad opportunities of the 21st century.

Chapter 8
Internationalizing and Elevating IIT (1996–2018)

Following the recommendation of the National Commission to expand IIT's international role and its image, I embarked on a trip to the Far East in early 1995. I was hosted first in Seoul, Korea, by my former doctoral student Chang Chi, then president of W. R. Grace Co., Korea. I also met Jong Soung Kimm, a former interim dean of our College of Architecture; he was head of SAC International, a prominent architecture firm in Korea. He opened doors for me, introducing me to leaders at several prominent universities in Korea.

In Seoul, I met S. R. Cho, who I had known as a master's student during my first year at IIT in 1964. Cho, who was chairman of the Hyosung Group, a conglomerate of approximately 31 companies, hosted a dinner and invited many of our alumni for me to meet.

After Seoul, I visited alumni in Bangkok and finally in Delhi, where I met with a large group of alumni. I was pleased to see that most of these prosperous alumni had fond memories of their teachers and mentors at IIT; at the same time, however, I was surprised to learn that they had not been in touch with IIT and had not heard from the university.

Upon my return, I reported my observations to Lew Collens. At the spring faculty meeting, I also made a presentation to the faculty, pointing out that with the help of successful alumni in various countries, especially in Asia, we should build international awareness of IIT.

Collens was supportive of the university's international outreach, and in 1996, the Office of International Affairs was officially established under my leadership. On July 1, 1996, Collens announced, "Effective today, Darsh Wasan will be the vice president for international affairs. As you know," he added, "Darsh Wasan has been concentrating his energies on our international programs. He has made considerable progress in building academic linkages with universities around the world, has visited frequently with Asian alumni to create a strong support base for the university, and has brought new international programs to the campus. As we move forward, Darsh will be able to concentrate his full energies on international affairs."

Mission and Vision for International Affairs

The mission of the Office of International Affairs was to develop relationships, strategies, and initiatives that advanced and extended globally the university's mission of teaching and research and enriching society. The vision was to propel IIT into becoming a renowned, diverse, and vibrant university that prepared its students to thrive in the global economy. It would be a university with *globality*, also attracting domestic students interested in adding international experience to their portfolios, as well as students and faculty from all over the world who sought a crucible in which to test their ideas and to innovate.

I implemented a multipronged strategy for internationalizing IIT that included actively engaging international alumni, establishing partnerships with universities abroad, implementing international distance learning, setting up IIT offices abroad, and forming an international board of overseers.

Over the past 20 years since 1996, the Office of International Affairs has made considerable progress. International enrollment grew from 15 percent of the total student body in 1996–97 to 49 percent in 2016–17 (see figure 8-1). Today, IIT has the highest percentage of international students among the Association of Independent Technological Universities (AITU) and is one of the highest in the nation. IIT now attracts students from 100 countries and has established global academic collaborative agreements with nearly 200 universities. IIT's Office of Digital Learning has delivered courses for several master's degree programs overseas via streamed video. More than 600 online students in India have completed their IIT master's degrees. IIT has established offices abroad in India, China, Europe, and Korea. In cooperation with a partner university, the Instituto Tecnológico y de Estudios Superiores de Monterrey, Mexico, IIT has secured a USAID grant of $500,000 to provide education and training in environmental management for Mexican students. During the last two major development campaigns (1998–2004 and 2010–16), the Office of International Affairs raised $6 million from international alumni and friends to establish an endowed chair professorship in chemical engineering and endowed graduate fellowships and undergraduate scholarships for international students, as well as other programs. And, the office has established the IIT International Board of Overseers, which includes global executives who advise IIT on global strategy.

Recruitment of International Students

To facilitate recruitment of international students and cooperation with foreign universities, IIT established centers in Bangalore (1998), Paris (1998), Beijing (2003), and Seoul (2005).

In the 2016–17 academic year, international students comprised 49 percent of IIT's total enrollment, and 55 percent of IIT's net tuition revenue came from international students. The two top countries of origin were India with 1,443 students, and mainland China with 1,427 students.

Illinois Institute of Technology (India) Private Ltd.

In December 1996, Motorola contacted me to see whether IIT could offer master's degree programs to its employees in Bangalore, known as India's Silicon Valley. The company wanted to give its top-performing employees the opportunity to advance their knowledge and skills in telecommunications and software engineering without leaving India.

IIT was already offering master's programs to U.S. suburban working professionals transmitting courses via microwave to company sites within a 40-mile radius of the Sears Tower (now the Willis Tower) in downtown Chicago. I seized the opportunity to extend the reach of IIT and to give something meaningful back to

India, my country of birth; I asked Motorola for seed funding of $100,000, which they provided.

I recruited D. "Raja" Rajagopalan, an electrical engineer who was a retired officer in the Indian Air Force, to manage the India operations. In 1997, we began offering a master's degree in computer science and a master's degree in telecommunications and software engineering in Bangalore. At first, 36 employees from five companies were enrolled. C. Robert Carlson, then computer science department chairman, went to India to teach the first course; it was a short course and delivered on-site, so that he could meet the students personally.

For the next few years, regular semester-long courses were offered by joining occasional live teleconferences with videotaped classes. We sent videotapes to Bangalore via air courier; the students submitted their homework via email or FTP sites. IIT/V Director Louise Hewitt and her staff were invaluable in this endeavor, managing all logistics and technical aspects of promptly delivering these courses.

In order to determine how best to operate in India, at the suggestion of good friend and IIT alumnus Sam Pitroda, I had contacted Nishith Desai, a lawyer and a chartered accountant in Mumbai. To offer our programs in India, Desai advised us to set up a private company: Illinois Institute of Technology (India) Private Ltd. He helped us apply to the Foreign Investment Promotion Board (FIPB) of India, and on August 4, 1998, the board gave us permission to operate. We were the very first and only U.S. university-approved institution in India to offer these educational programs. (See Appendix 1 for a more detailed account of how we got started in India by D. "Raja" Rajagopalan.)

We are proud of our India program, for it has helped us not only build greater awareness of IIT and better serve students and their employers, but also because the students have been indeed grateful for this opportunity. The companies sponsoring the students told us that, at a time of high turnover, this initiative helped them retain their highly skilled employees. Moreover, some companies even commercialized the ideas that their employees had begun to develop as projects in their IIT classes.

In 2000, when Internet bandwidth and its reliability in India made it possible for us to deliver courses via streamed video, we expanded our programs and locations. We now offer 13 graduate-degree programs in six cities: Bangalore, Chennai, Hyderabad, Mumbai, Pune, and Delhi. Because we offer the same rigorous courses that we do in Chicago, IIT is well respected there. In fact, some students now choose to complete their study program partly in India and partly on our Chicago campus.

Offering courses in Bangalore has helped us spark a dramatic increase in enrollment of Indian students at IIT. In 1997, for example, only 250 students from India were enrolled on our campus; within a few years, their enrollment more than doubled. By 2002, IIT was the No. 1 destination for Indian graduate students applying for U.S. visas. Indian students not only enhanced the multicultural vibrancy of our campus and added revenue to IIT, they also contributed professionally to the U.S. and to India.

The China Office

I first visited Beijing in 1996 and met with Jeffrey Li, an IIT alumnus, who used to manage the China operations for a company owned by Bob Pritzker. I asked Li for advice regarding IIT having a presence in China, so that we also could directly recruit students from China. It took several years and a few trips to China before we were able to establish an office in Beijing in 2003. Tim Zhang, who received his M.B.A. from the Booth School of Business at the University of Chicago, was the founding director of the IIT China Office. (See Appendix 2 for Zhang's account of the starting the IIT operation in China.)

Today, the IIT China Office, directed by Hua Qu, monitors independent student-recruiting agents; represents IIT at educational fairs; assists prospective and admitted students; holds orientations for admitted students; connects IIT with alumni in China; and facilitates connections for IIT faculty and staff who are visiting China.

The Paris Office

In 1997, Vanita Misquita, an IIT alumna and former director of the IIT International Center for international-student and scholar services, informed me that she would be leaving for France to marry a Frenchman. In order to create the university's presence in Europe and to develop partnerships with reputable schools there, I discussed with her the idea of setting up an IIT Paris Office. Coincidently, our College of Architecture was looking for a place for its semester study- abroad program. In 1998, Misquita registered an IIT office in Paris so that the College of Architecture could rent a studio there, but rather than renting her own IIT office on the premises at a high cost, she chose to operate from her own apartment in Paris. (See Appendix 3 for Misquita's brief account of how we got started.)

Misquita's not only established dual-degree master's program alliances with several European universities, she also set up a visiting-student program for undergraduates from Europe, a nondegree research scholar program for European students that provided numerous study-abroad opportunities for IIT students as well. She has been recruiting more than 150 IIT students per year.

In 2014, our College of Architecture closed its studio in Paris when the new dean of architecture decided to move the office to the Netherlands as he already had his business office there.

The Seoul Office

IIT's office in Seoul, established in 2005 and headed by S. H. Noh, brings visiting students and dual-degree bachelor's students to IIT, as well as cohorts of certificate students to IIT's Center for Professional Development for short-term programs. It also cultivates relationships with various universities and high schools in Korea. Noh works very closely with our alumni in Seoul, where we have a very active alumni chapter.

International Board of Overseers

In 2001 and at the suggestion of the newly elected trustee Kaarina Koskenalusta, who at the time was president and CEO of the Executives Club of Chicago, IIT established an International Board of Overseers; its membership consisted of global executives.

It was chaired first by Claude Smadja, then managing director of the World Economic Forum in Davos, Switzerland, and then by Göran Lindahl, former ABB president and CEO and a director of Sony Corporation and IKEA Greentech. The board met twice a year, once on the IIT campus and a second abroad.

The meetings abroad were held in various cities, such as London, Paris, Beijing, Mumbai, and Prague. Board members included Christine Lagarde, now managing director of the International Monetary Fund; John W. Rowe, chairman and chief executive officer of the Exelon Corporation, who later became chairman of the IIT Board of Trustees; and Marshall M. Bouton, then president of the Chicago Council on Global Affairs; as well as IIT alumni including Sam Pitroda, prominent Indian business leader and then president of World Telecom; Jamshyd N. Godrej, an Indian industrialist and current managing director and chairman of Godrej & Boyce; S. R. Cho, then chairman of the Hyosung Group of Korea; and Ajva Taulananda, then president of Telecom Asia of Thailand.

The board advised the IIT administration on developing a global strategy and brand for the school, building alliances that would advance the university as a whole, and acquiring resources. It should be noted that this board has been inactive since 2011, when its chairman Lindahl resigned as a trustee.

Engaging International Alumni

It has been one of my key initiatives to engage our international alumni and to foster networking. Our global centers in India, Korea, China, and Europe continue to be actively involved in reaching out to our alumni in those countries. Our efforts have included identifying nearly 100 executive-level donors and engaging them in a scholarship campaign; recruiting them to serve on the advisory committees of the university; and electing them to serve as university trustees (S. R. Cho and Sam Pitroda, for example, served their three terms as university trustees; Madhavan K. Nayar, who established the Nayar Prize, currently serves as a trustee; and Jamshyd Godrej completed his three terms and now serves as a lifetime trustee). These executives have helped us establish IIT Alumni Association chapters in Korea, Hong Kong, Thailand, Taiwan, Japan, India, and China and bring together representatives about once every 18 months for a Pan-Asian Alumni Association get-together.

The first meeting of our Asian alumni took place in Hong Kong in 2000. IIT Institute of Design alumnus Victor Lo Chung-Wing, chairman and CEO of Gold Peak Industries and currently an IIT trustee, was the host. Both Victor Lo and Cho agreed to serve as honorary chairs of this gathering and helped me organize subsequent meetings in Bangkok, Seoul, Tokyo, Mumbai, and Beijing.

Then IIT President Lew Collens and his wife, the late Marge Collens, attended the alumni gatherings in Bangkok and Seoul; then President John Anderson attended the meetings in Mumbai and Beijing; and more recently, current IIT President Alan Cramb has attended the meetings in Seoul and Paris.

In 2014, the IIT Office of Alumni Relations, under the directorship of Jamie Acton, organized the first global alumni gathering on the IIT campus and coordinated the most recent global gathering in Paris in 2016, which was locally organized by Vanita Misquita and also the newly founded European Alumni Association, headed by Charlotte Leroy.

Establishing Endowed Hyosung S. R. Cho Professorship

I knew S. R. Cho when he was a graduate student in the chemical engineering department at IIT. I had joined the department as a young faculty member in 1964 when there were significantly fewer international students.

Cho had dreamt of becoming a college professor, but in 1965 he was called upon by his father, the founder of Hyosung Trading Company, to return to Korea. His father had established a manufacturing company and had planned to build one of the most advanced nylon fiber production facilities in the world. He wanted his chemical engineer son to join Hyosung's management and oversee the construction of the facility.

In the 1970s, Cho played a key role in the growth of Hyosung; he not only led the construction of the company's nylon plant, he also helped establish several other companies. He led the Hyosung Group as CEO from 1981 until 2017, when he retired; his son, Cho Hyun-Joon, is the current CEO.

Hyosung has fueled Korea's economic growth; it consists of 23 manufacturing corporations, 8 trade corporations, and 19 trade offices around the world. Its businesses provide technology and services for various industries, ranging from power to industrial materials, fibers, chemical construction, trading, and information technology. Its products include tire cords, PET bottles, electrical transformers, spandex, and technology for ATM machines.

Cho is highly regarded in the Korean business community for his expertise, knowledge, and global insight. In March 2007, he was appointed chairman of the Federation of Korean Industries. He also has served as a vice chair of the Korea Chinese Business Association, the Korean chair of the U.S.-Korea Business Council, and the chair of Japan-Korea Business Conference and has been a proponent of Korea forming a free trade agreement with Japan in addition to the U.S., the E.U., and China.

Thirty years after Cho returned to Korea, when I was provost and exploring how to globalize IIT, I flew to Asia to visit some of our prominent alumni. My first stop was Seoul where I met with Cho. He organized and hosted an event for me, introducing me to some 70 alumni. I found out he was funding all the activities for the Korean alumni, and Korea became IIT's first official alumni association chapter abroad.

When I returned from this trip, my Office of International Affairs established the Distinguished Asian Leadership Awards to honor such international involvement and support from alumni. At the inaugural international gala in the fall of 1995, we presented the awards to Cho and three other prominent alumni leaders: Ajva Taulananda and Dhongchai Lamsam, then president of Loxley Public Company, both from Thailand, and Sam Pitroda, from India. Today, the IIT Alumni Association honors this type of service to IIT with the International Award of Merit.

When I met Cho in Seoul, I learned that he had never received his master's degree from the university because his father had called him home before he could complete the required research thesis. When I returned to Chicago, I looked up his academic record in the registrar's office, and I found out that he had spent four years at the university. He had taken a lot more courses than required at the time for the master's degree. I knew that several years after he left IIT, the university began to award professional master's degrees based on course work alone. To me, it looked as though Cho qualified for the professional master's degree as he had maintained the required GPA and had sufficient credits. I discussed his academic record with the dean of our Graduate College, as well as with the chair of the chemical engineering department. They concurred. I decided to surprise Cho with the awarding of his master's degree at the time of the international gala dinner. When Lew Collens presented him with his diploma that evening, Cho was so moved that tears welled up in his eyes. He was totally surprised and delighted. Over the years, he mentioned to me how pleased he was to finally have his diploma.

In 1996, Cho was appointed to the IIT Board of Trustees; he was the first international trustee. He asked me then what he could do for IIT since he could not attend all the trustee meetings held on the Chicago campus.

On December 23, 1999, two days before Christmas, I received a call from Cho. He spoke about his great education at IIT and the great influence IIT teachers have had on students. It appeared that Cho really wanted to inspire and encourage the great professors at IIT. At the time, IIT had its Pritzker-Galvin Challenge grant. I told him that we would like to see a chaired professorship named after him and that he could double his impact, since his gift would be matched by the challenge grant. He made a gift of $1 million for the professorship and $500,000 for a graduate fellowship to IIT. I was so excited that I called Lew Collens, who was vacationing in Hawaii, and gave him the good news.

This was when I also asked Cho if he could help us form an Asian Alumni Association to connect IIT alumni throughout Asia, so that they could network with each other and when he and Victor Lo agreed to serve as honorary chairs of our first Asian Alumni Association meeting held in Hong Kong in 2000. We have been meeting about every two years since then.

Cho also served on our International Board of Overseers and attended several regular meetings of this board. In October 2012, he sponsored and attended the Asian Alumni Association meeting in Beijing. At that meeting, we also formally established an alumni chapter in China and held an investiture ceremony for Vijay Ramani, the Hyosung S. R. Cho Endowed Chair Professor for Chemical Engineering. In attendance were IIT's president, provost, two deans, two vice presidents, and several professors and staff members, as well as 150 alumni.

In addition to his gifts, Cho has continued to fund activities for the Korean alumni; he also contributes regularly to the Korean Students Association at IIT. When the World Cup was held in Seoul, he even took some alumni from Korea and Thailand to see a soccer match.

Cho helped me make connections and establish cooperation with several prominent Korean universities. Today at IIT, after China and India, students from Korea make up the third largest number of international students.

Cho may not have become a professor as he once dreamt, but he has helped IIT become a global university, for which we are most grateful to him.

In 2013, Cho received an honorary doctorate from IIT, and he donated another $1 million to the university.

Taking IIT to the Next Level

IIT now has had decades of experience offering interprofessional project-based learning. We have a large international student body, a Leadership Academy, and the Jules F.

Knapp Entrepreneurship Center. More recently, the Lewis College of Human Sciences began to offer a global studies program, and the Ed Kaplan Family Institute for Innovation and Tech Entrepreneurship, a 70,000-square-foot building that fosters active and collaborative learning, opened in October 2018. These initiatives help the university increase student diversity, build research strength, and better prepare students for the global economy, helping IIT to develop an overall cohesive global brand.

To further integrate and capitalize on these strengths, in 2008, I proposed the establishment of a new interprofessional, international residential college to create a distinctive honors college that would draw a broader base of well-prepared and high-achieving students from around the globe. These students would be able to afford tuition and would attract philanthropic and industrial support that would fund students who could not otherwise afford tuition. Utilizing the strength of IIT's global student body and the opportunities afforded by Chicago as a global city, I proposed to create a unique living environment in which to learn and to innovate together, focusing on solutions to global problems using our interprofessional project structure.

U.S.-born and international students and undergraduate and graduate students would together reside in such a college. They would be chosen on the basis of quality, potential for leadership, and international awareness. The establishment of this college would differentiate IIT from other technology-focused universities, strengthening the bond between IIT and its international education partners to address issues of global concern, such as energy, environment, water, health, and cybersecurity.

The proposed college would give its students' knowledge, skills, and a global perspective and mindset that would complement their specific academic expertise. The main elements of this new initiative would include a residential experience; language resources; an international experience and courses to enhance diverse cultural awareness; community service; leadership, team, entrepreneurship, and ethics training; and interprofessional projects with cooperating universities abroad to address major problems of global concern.

To support the establishment of this new residential college, funds would need to be raised in endowed and capital dollars. Funding would be required for a new residential building, endowed undergraduate fellowships for study abroad and global projects, endowed graduate scholars' fellowships, the hiring of a director and staff, a travel fund for faculty, an annual symposium for cooperating universities, and phase one programming and planning.

My hope is that a new interprofessional, international residential college will be one of IIT's future endeavors.

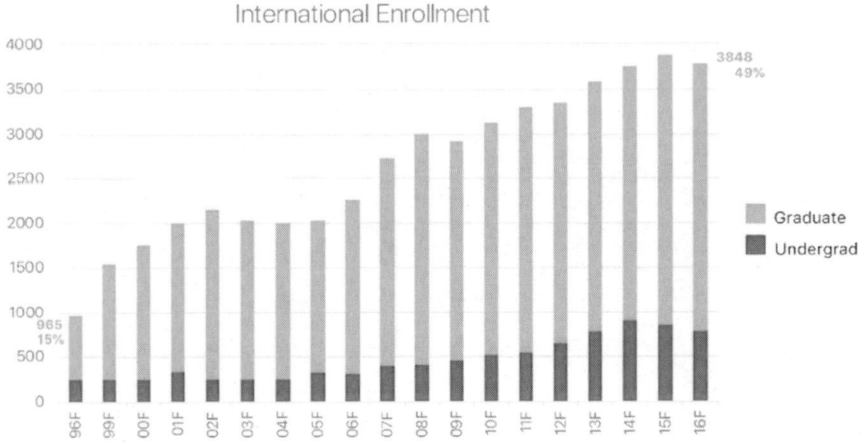

FG-8-1 IIT's international enrollment from FY1996 to FY2016

2012 Beijing

2014 Chicago

2013 Seoul

International reunions

Chapter 9 - Applications of My Research

My research activities over the past 50 years have focused on dispersion science and engineering, including fine-particle dispersions, foams, emulsions, and the importance of the liquid interface between the liquid-gas, liquid-liquid, and liquid-solid surfaces in dispersed systems. Dispersion science holds the key to cost-effective manufacturing of a wide variety of chemical, petroleum, pharmaceutical, food, and consumer products.

My research teams have changed over the years as students and professors have joined IIT and then moved on; about 125 M.S. and Ph.D. students and 12 postdoctoral fellows have worked with me. Our teamwork has produced a total of 450 publications, including 30 chapters in research monographs; six patents; and a textbook, *Interfacial Transport Processes and Rheology* (Butterworth-Heinemann, 1991), coauthored with David A. Edwards, my former doctoral student and now a professor of industrial practice at Harvard, and the late MIT professor Howard

Brenner. Our teamwork has also culminated in research dissertations of over 120 students. (See figure 9-1 for a list of the master's and Ph.D. students I have supervised over the years.)

Financial support for my research endeavors was not only provided by federal agencies such as the National Science Foundation (NSF) and the Department of Energy (DOE), but also by more than 30 industrial organizations, some of which are mentioned in earlier chapters. What is most remarkable is that our research has secured the competitive financial support of the NSF for more than 40 years without any disruption. I am grateful for this. The total extramural funding for my research over the past 50 years is estimated to be nearly $25 million, and during my tenure as a principal investigator and coinvestigator, I have secured for IIT more than $100 million in funding from federal agencies and industry.

As an engineering educator, my research philosophy has always been to work on problems that *need* to be solved rather than seeking problems that *could* be solved. Most of the research in academia is of a fundamental nature. From the beginning of my career, however, I have guided my research groups toward solving real-world problems encountered by industry; I conducted studies in full collaboration and consultation with members of the industrial-scientific community.

I have maintained good ties with industry and academic researchers in the U.S. and in Eastern Europe (especially Bulgaria) even in times when such collaborations were difficult to maintain. As I mentioned in Chapter 5, regarding the birth of the NCFST, I was fortunate to be one of the first academics in chemical engineering to promote the concept of joint industry- university research programs.

Industrial Technical Advisory Committee

In the 1970s, I set up the Industrial Technical Advisory Committee (ITAC) at IIT to provide direction for my basic research program. It was comprised of senior scientists and research directors from about 20 companies, including two European companies, developing chemical, petroleum, and consumer products. My research group of 10 to 12 students, a few postdocs, and I met with them annually at IIT.

We presented the results of our research at daylong meetings, followed by a lab visit to show industry professionals our novel experimental methods. The day ended by soliciting, in writing, comments on the quality of our program, as well as on the topics that they felt were of interest to them and to their organizations. We asked them to identify key elements that would improve our research program; they were all required to submit their comments before leaving the campus. I would later discuss their comments with my group and plan our future research. Over the years, my teams have helped solve problems related to the environment, energy, consumer products, food, and health. A few of these accomplishments are highlighted below.

Remediation of Nuclear Waste

The processing of nuclear materials produces radioactive waste hazardous to human, plant, and animal life. The waste is stored in large underground tanks in radioactive immobilization facilities, or waste-treatment plants, such as the DOE sites at the

Savannah River in South Carolina and Hanford in Washington State. The Savannah River site, built in the 1950s, is a repository for about 42 million gallons of waste from the development of nuclear weapons. The Hanford site has been in use since World War II, when the government began to store waste from the Manhattan Project; the site contains about 56 million gallons of radioactive waste.

In 1987, Hanford was closed to further waste deposition; however, a serious problem emerged when some of the storage tanks developed leaks due to corrosion. In the early 1980s, it was recognized that there would be significant cost and safety advantages associated with immobilizing the radioactive waste in a stable, solid form. Immobilization would keep the waste from contaminating the water table and the surrounding environment.

The EPA designated vitrification (converting a substance into glass or a glass-like end product, usually by heat) as one of the available and accepted technologies for the immobilization of highly active nuclear wastes. The wastes can be immobilized by incorporating them into a borosilicate glass matrix, thus reducing the mobility of the radionuclides, the radioactive forms of elements.

By storing the wastes as solids, it would be possible to allow for their safe, eventual decay. Much of the Savannah River site's high level of radioactive waste was in the form of caustic suspensions of insoluble, inorganic (sludge) particulates. The chemical processing of this sludge to produce melter feed suitable for a glass melter involved boiling the sludge.

It was at this stage that a critical problem arose. Boiling the sludge would create severe foaming that would cause excessive overflow of radionuclides and radioactive waste and lead to serious health and safety issues. Foaming could also severely impact the processing rate and the cost of canister production.

A thorough understanding of foaming and a suitable antifoamer are essential to creating an effective process for sludge storage. When most people think of foams, they think of soap or shampoo. However, the foam that develops when nuclear waste is boiled is different. Foaming during the evaporation of this type of nuclear waste consists of three phases: gas, liquid, and finely divided solids; the foams aggravated by the particles encountered during evaporation are different from those formed by common surfactants (as in shampoo). Three-phase foams that incorporate finely divided solids (colloidal particles) are significantly different from common foams, making the challenge of a safe boiling process even greater.

I am pleased to say that my research team tackled the challenge of helping to safely store nuclear waste with great enthusiasm. In our studies, supported by a grant from the DOE in the mid-1990s, we addressed two different but interrelated problems. The first dealt with the fundamental understanding of the causes of the foaming and foam stability. The second addressed the need for a practical and effective antifoam or antifoaming agent to minimize the production of foam during the processing of the nuclear waste.

Typical commercial antifoams were found to be ineffective, because they were unstable in the harsh environment created by the waste material. Radioactive waste processing includes strong caustic or acidic solutions, high temperatures, and radiation. Commercial antifoams that eliminate foaminess in the presence of surfactants are not effective with foams that are stabilized by fine particles.

The commercial antifoam agent being considered for use (Dow Corning 544) was found to be ineffective under boiling conditions. With our mechanistic understanding of foam generation and its stability, we then developed a novel antifoam, called IIT-747. Alex Nikolov created it in our laboratory using simulated (nonradioactive) waste. It minimizes foam formation in ultrafine particle-laden slurries during high-radioactive-level waste concentrations prior to vitrification. One doctoral and three master's students carried out the fundamental studies, funded by the DOE, for their research dissertations. The antifoam was subsequently tested in a pilot plant at the Savannah River site with real waste in their shielded cells.

It was a great success. Since 2008, IIT-747 has been helping safely process nuclear waste at the Savannah River site. Details of our new antifoam formulation were published in well- known journals, such as the *Journal of Environmental Engineering*, to further advance the study of waste remediation. The formulation is also available on the federal government's Office of Scientific and Technical Information website.

Clean Synthetic Fuels

Many of the fuels used today can cause environmental and health issues, both directly and indirectly. During the 1970s and 1980s, our team worked on developing processes for separating suspended fine particles from nonaqueous media to produce clean fuels. Synthetic fuels derived from coal, shale, and tar sands contain particles of unreacted solids (ash- carbonaceous residue), which impede downstream processing. The complex nature of the colloid chemistry in such nonaqueous media makes the removal of these particles difficult.

My research team relished the opportunity to solve this thorny problem. We first developed a method to remove the fine particles suspended in nonaqueous media using a high- voltage electric field. With the generous funding from Amoco Corporation, the NSF, and the DOE, Dimitri Gidaspow and I invented two practical devices based on electrokinetic phenomena: a cross-flow electrofilter and a lamella electrosettler for the separation of colloidal particles from nonaqueous solvents. This led to two U.S. patents. Our methods, which were more energy-efficient than conventional techniques, were used to upgrade synthetic crudes and other organic liquid slurries. In 1986, the NSF gave me and Gidaspow a Special Creativity Award for electrokinetic phenomena in nonaqueous media.

Cool Whip

Our research team also investigated how to improve *Cool Whip* and other whipped toppings. These are food products produced by dispersing about 600 to 700 percent (by volume) air into an aqueous, creamy phase. They contain hydrogenated vegetable oil, including coconut and palm oils; high fructose corn syrup; skim milk; thickeners, such as xanthan and guar gums; emulsifiers, such as caseinate, a protein derived from milk, and monostearate; and beta carotene for coloring. Air is incorporated into a cream mixture containing at least 30 percent (by weight) of fat particles, by either mechanical high-speed blade agitation or by injecting a gas under high pressure, which expands rapidly when released from a pressurized container. The incorporation of air into a food product imparts lightness and improves its texture and mouth-feel.

During air whipping, the viscosity of the product increases with the increase in the volume of air. The fat particles then form a network structure (that is, they aggregate) trapping gas bubbles. If, however, whipping continues, the fat particle texture will be disrupted, resulting in the collapse of the whipped product. The texture and stability of such foam-based products are therefore very important. The development of a stable structure (and suitable aerated food product) depends on the interactions between the fat globules and the air bubbles under the shear stress conditions that exist during whipping.

In the early 1990s, the Kraft Foods Group (which was purchased in 2015 by H. J. Heinz Company and became the Kraft Heinz Company, the fifth largest food company in the world) approached us to undertake a basic research study to develop a comprehensive understanding of *Cool Whip*'s dispersion structure, texture, and stability utilizing the novel instrumentation and methods developed in our colloid and interface science laboratory at IIT.

I assigned this project to Kalman Koczo, a senior research associate in my lab. Koczo was from Hungary and had a B.S. in chemical engineering, though his doctorate was in physical chemistry. He had been working with me on a related project, investigating oil-foam interactions, such as those that occur when shampooing hair. Together with Alex Nikolov and a couple of my doctoral students, we undertook a study on whipped toppings, funded by Kraft Foods for nearly a decade.

We not only established the key mechanisms that degrade the aerated product, such as the gas diffusion between bubbles, but we also measured the bubble size distribution and its effect on its viscosity and structure. We used the light scattering (the Kossel diffraction) method to monitor the fat particle aggregation structure formation and its variation under shear stress conditions (whipping).

Working with two of senior scientists from the company, we developed a comprehensive mechanistic model for *Cool Whip*. It was subsequently tested in the pilot plant at Kraft Foods under different temperature and pressure conditions suitable for product transportation and its storage. The company later developed the patented technology for this product.

Blood Substitute

The American Red Cross has estimated that one person in the United States requires a transfusion every two seconds, producing a critical need for a disease-free, universally transfusable, and oxygen-carrying blood substitute for the emergency resuscitation of trauma victims, as well as for high blood-loss surgical procedures. There are, of course, many other applications for the use of artificial red blood cells, such as treating patients suffering from aplastic anemia or sickle cell anemia. (A blood substitute can also be used for veterinary medicine, where blood banking is not practiced.)

Currently, there are no perfect substitutes for human blood because the entire human body, including blood, is complex. The properties of the ideal blood replacement must include high oxygen-carrying capacity, no toxic effects, histocompatibility (the ability to share tissues without causing a negative immune response), and adequate storage stability. None of the available blood substitutes have satisfied all these criteria.

Richard Beissinger and I were contacted by a team of surgeons at Michael Reese Hospital, then located near the IIT campus. The physicians there suggested we consider developing artificial blood using, as a starting material, a concentrated hemoglobin solution, the iron- containing, oxygen-transporting metalloprotein in red blood cells.

Working with a couple of our graduate students, we proposed an artificial blood substitute created from multiple emulsions, which are complex polydisperse systems. The substitute consisted of a concentrated hemoglobin solution encapsulated in an oil phase, which was then dispersed in an isotonic saline solution. The multiple emulsion droplets essentially mimicked red blood cells in that, in both phases, the hemoglobin solution is the encapsulated phase.

The oil phase in the multiple emulsion served as the membrane across which oxygen and carbon dioxide gases were exchanged. In this methodology, a concentrated hemoglobin solution was emulsified in vegetable oil to form microdroplets by using a nontoxic surfactant in the oil phase. This was followed by the dispersion of the primary emulsion into an aqueous (buffer) phase containing another surfactant to form the multiple emulsion product.

Beissinger created a model multiple-emulsion and devised quantitative methods to assess its yield. The yield tests measured the amount of hemoglobin released into the external phase, the drop-size distribution, oxygen-carrying capacity, short-term storage stability, and viscosity. The yield obtained for the formulated product was excellent. It had a high oxygen-carrying capacity, an average diameter of the prepared droplets of eight micrometers (thus comparable to the size of actual red blood cells); also, a high yield could be maintained for four days at 4°C. While many ideas work well in theory, they do not always pan out in practice; so, we decided to put our ideas to the test. We carried out both in vitro and in vivo tests using this multiple emulsion formulation by providing transfusions to rats. After receiving the blood substitute, the rats survived for over 24 hours, which was a positive sign but not quite the solution we had hoped for. Moreover, the long-term stability of the multiple-emulsion product was inadequate. To this day, finding a blood substitute remains an issue that must be solved.

Although this study did not provide all the answers to the questions regarding a blood substitute, it was, nevertheless, a step in the right direction. Our work was reported in a number of professional journals and resulted in two U.S. patents. The project was not pursued further because Beissinger fell ill and sadly passed away in 2003.

Nanofluid Applications

Nanotechnologies, once the stuff of science fiction, have now become an important part of our arsenal in developing new solutions to problems. Nanotechnologies, such as nanofluids, which are comprised of liquid suspensions of nanoparticles, are used in a wide variety of technological applications and biological systems. Their spreading and adhesion behavior on solid surfaces, for example, can yield new materials with such desirable structural and optical properties as solar cells and photonic devices. The spreading behavior of nanofluids also has practical implications for soil remediation, oily soil removal, enhanced oil recovery, detergency, and hydraulic fracturing of gas and oil to enhance well productivity.

In a paper published on May 8, 2003, in *Nature*, Alex Nikolov and I reported on the mechanism that could enhance the spreading of a particle-containing fluid on a solid surface. According to our mechanism, when a fluid containing tiny, uniform spheres spreads across an oil drop, the spheres begin to concentrate and organize themselves, packing tightly into what effectively serves as a wedge between the sphere and solid surface (see figure 9-2). As a result, the rest of the fluid now has a lower concentration of nanospheres than does the wedge and will naturally flow toward it to equalize the concentration. This flow, in combination with the wedge structure, lifts the oil drop and the fluid flows underneath it.

We used video microscopy and directly observed that a crystal-like structure occupied the space where the oil drop had met the solid surface. The nanoparticle structuring phenomenon gives rise to the structural disjoining pressure (a force normal to the interface) in the wedge, thereby enhancing the spreading of the nanofluid on the solid surface, detaching the oil drop.

The driving force for the enhanced spreading of the nanofluid is the oscillatory structural disjoining pressure (a high osmotic pressure). The disjoining pressure is high near the vertex (see figure 9-3) because of nanoparticle structuring in the wedge confinement. As the structural disjoining pressure increases toward the vertex of the wedge, it drives the nanofluid to spread at the wedge tip. This behavior enhances the spreading of the nanofluid, eventually detaching the oil drop.

This important discovery has led to better solutions for everyday problems. Over the last decade, for example, we have explored several practical applications of this new mechanism, as described in the following sections on faster-acting detergents, enhanced oil recovery, and bacterial removal and inactivation from fresh produce.

Faster-acting Detergents

As a young boy growing up in India, I used to watch my grandmother washing her laundry by first wetting it with water containing only a tiny amount of soap and then striking the wet fabric against a rock. I often wondered how she could clean the laundry without using more soap, though with abundant elbow grease! Years later, my grandmother's efforts became my inspiration when I began working with nanotechnology. Nikolov and I decided to explore this mechanism behind her way of soil cleaning. We used ground stone (inorganic silica dioxide nanoparticles) to replace the detergents for removing oil from a soiled surface.

After the publication of our paper in *Nature*, Procter & Gamble Company took an interest in our work. The company funded our research to apply our new mechanism using an oil sample, cotton fibers, and its *Tide* detergent. We found that the IIT silica nanofluid performed very well in lifting an oil drop based on this new mechanism, and the results compared favorably with those by using the *Tide* detergent.

Our research clearly showed the potential application of our new mechanism for developing environmentally friendly green cleaners by replacing environmentally unfriendly detergents or surfactants derived from petroleum products. Procter & Gamble, along with IIT, filed for a U.S. patent.

Enhanced Oil Recovery

While it is important to look for alternative energy sources, it is equally vital to realize an efficient approach to using what we already have. Frac Tech Services, now FTS International, an American company that specializes in providing enhanced oil and gas recovery solutions, supported us on a second exploratory project based on our new mechanism.

We tested the performance of our hydrophilic (water wetting) silica dioxide particles, using Texas light crude oil and a reservoir brine solution (with a low salinity formation in water) at 25°C against a microemulsion (a highly concentrated complex surfactant mixture) additive used in industry. The results were outstanding; the IIT nanofluid formulation utilizing our novel mechanism outperformed the microemulsion additive by 20 percent in crude oil recovery. I am pleased to report that FTS, along with Dr. Holcomb, and Nikolov and I, filed a patent application in 2009. FTS has subsequently explored the use of nanoparticle dispersions in other treatments as well, such as paraffin, asphaltene, and slurry removal, and the recovery of hydrocarbon from tar sands. The company has commercialized the use of nanoparticle dispersions based on our new mechanism.

Other companies also showed an interest in applying our research to solve their challenges. There was, however, one aspect holding back our new nanofluid from being used in certain oil-recovery situations. The IIT silica nanofluid could be used for oil recovery when a petroleum reservoir contained a low-salinity brine solution—that is, when its formation water contained a low amount of salts and only monovalent ions, such as sodium or potassium chloride. Ours was *not* suitable for a high-salinity brine environment, but that was about to change.

In 2012, Chevron provided us some funding so that we could further explore the use of our novel mechanism. The goal was to recover oil from a petroleum reservoir with a high-salinity brine solution, especially one that contained calcium and magnesium ions. With persistence, we finally succeeded in developing a new IIT polymer nanofluid formulation with the potential working under high-saline conditions.

We performed oil-recovery tests on the crude oil and rock samples supplied by Chevron using the new IIT polymer nanofluid and a harsh brine solution at a reservoir temperature of 55^{0}C. We were pleased to find out that our nanofluid excelled in the harsh brine saline environment. The results of our work were published in 2014 in the journal *Energy & Fuels*.

There is, of course, still much work to be done in this area. Beginning in 2014, Schlumberger-Doll Research Center began funding our research using the new IIT polymer nanofluid for recovering oil from petroleum reservoirs, including one with fractured rocks. The company, along with IIT researchers, filed for two patents. A former doctoral student who worked on this project in our lab is now employed by Schlumberger, and a second student, now working closely with the company, will soon complete his doctoral dissertation.

Bacterial Removal and Inactivation from Fresh Produce

Anyone who has suffered from a food-borne pathogen knows well that the illness and its recovery are indeed challenging. The World Health Organization has estimated that every year one out of ten people becomes ill from contaminated food. Some of the worst offenders are leafy greens, which can harbor dangerous bacteria. Preparing food as cleanly as possible not only supports the good health of a person, but it also helps the community, the farmers, and various businesses that participate in providing the freshest foods possible. Our new mechanism was able to provide a novel approach to the removal and inactivation or, more simply, the killing of bacteria using a unique nanofluid formulation.

Perhaps the most important part of this research has been our ability to understand and develop a clear explanation for what was happening. What, for example, caused the interaction between bacteria and solid surfaces resulting in their adhesion? We knew well that removing bacteria was vital to clean food, but deactivating bacteria was equally important. Because we were dealing with products for human consumption, we needed to develop something that was safe and nontoxic.

A combination of an organic acid, such as levulinic acid, and a nanofluid comprised of a SDS (sodium dodecyl sulphate) micellar solution was effective for removing and inactivating *E. coli* (which can cause organ failure and death) from a hydrophobic material. However, the mechanisms for enhanced microbial removal and inactivation using this combined chemical formulation were not well understood. We decided to investigate using *E. coli* K12 (obtained from a supplier) on a lettuce leaf (obtained from a grocery store in Chicago). We utilized the cutting- edge microbiological laboratory at IIT's Institute for Food Safety and Health (IFSH).

Using atomic force microscopy, we first determined that the disjoining structural pressure for the SDS micellar solution was rather high. Fortunately, however, this was the right kind of result, for it meant that our new mechanism could be used to remove the bacteria. Since we had to determine the role of the organic acid, we measured the electrophoretic mobility of the bacterium at different concentrations, i.e., various pH values. We found that at a pH of 2.6 (an acidic pH somewhat like that of orange juice or vinegar), the electrical charge on the bacterium was neutralized; the bacterium lost its charge completely. We found out that it was easy to remove and kill the bacteria on the lettuce leaf using a combination of 0.06 M SDS and 0.25 M luv acid, at a pH of 2.6. We used video microscopy to verify that our mechanism was indeed working. This research resulted in a mechanism useful to industry to develop an efficient bacterial sanitation formulation.

Research Recognition

There is great satisfaction that comes from doing your job well, as well as from discovering novel and useful solutions to practical problems. Still, I am most proud of the fact that my research students achieved notable accomplishments of their own after graduating from IIT and that many of them have become recognized leaders in their professions. The work performed by my group has also been recognized with numerous awards and honors bestowed by professional societies and other organizations. (See figure 9-4 for a short list of some of these awards and honors.)

I am also so honored to have served for 22 years as the editor-in-chief of a premier international research journal, *the Journal of Colloid and Interface Science,* and the journal's publishing company, Elsevier, has established the annual JCIS-Darsh Wasan Award to recognize individuals with outstanding contributions within the surface and colloid chemistry.

No.	Name	Degree
1	Aderangi	PhD
2	Agrawal	MS
3	Alexander	PhD
4	Baid	MS
5	Baker	PhD
6	Bindal	MS
7	Borwankar, C.	MS
8	Borwankar, R.	PhD
9	Bouillard	PhD
10	Brady	MS
11	Brauer	MS
12	Caceres	MS
13	Chad	MS
14	Chan, C.	MS
15	Chan, M.	MS
16	Chang	MS
17	Chatterjee	PhD
18	Chen	MS
19	Chengara	PhD
20	Chi, C.	PhD
21	Chi, V.	PhD
22	Cho	PhD
23	Chowdiah	PhD
24	Ciempil	MS
25	Chung	PhD
26	Cohn	MS
27	DiNardo	MS
28	Djabbarah	PhD
29	Dutta	MS
30	Edwards	PhD
31	Faghihi-Shirazi	MS
32	Goyal	MS
33	Gupta, L.	MS
34	Gupta, R.	PhD
35	Han	PhD
36	Hirschhorn	MS
37	Horiuchi	PhD
38	Huang	PhD
39	Israel	MS
40	Jakhete	MS
41	Jang	PhD
42	Jayaswal	PhD
43	Jones	PhD
44	Kaellis	PhD
45	Kao	PhD
46	Kim	PhD
47	Knapik	MS
48	Kondiparty	PhD
49	Kong	PhD
50	Krawczyk	PhD
51	Kumar	MS
52	Kurzeja	MS
53	Kush	MS
54	Lee, C.	PhD
55	Lee, J.	PhD
56	LI	PhD
57	Lim	PhD
58	Liu	PhD
59	Lo	PhD
60	Lobo	PhD
61	Lynch	MS
62	Mahajan	MS
63	Malhorta	PhD
64	Malviya	MS
65	Mansoori	MS
66	Maru	PhD
67	Menon	PhD
68	Milos	MS
69	Mukherjee	PhD
70	Nagarajan	PhD
71	Naik	PhD
72	Nevrekar	MS
73	Panitz	MS
74	Pasquarelli	MS
75	Patel	MS
76	Perl	PhD
77	Pfeiffer	MS
78	Pintar	PhD
79	Punwani	MS
80	Ramakrishnan	PhD
81	Ranade	PhD
82	Randhava	PhD
83	Randle	MS
84	Rao	MS
85	Rasouli	PhD
86	Rosenfeld	PhD
87	Rudin	PhD
88	Sabnis	MS
89	Sampath	MS
90	Sankar	MS
91	Schmidt	MS
92	Sethumadhavan	PhD
93	Shah, A.	MS
94	Shah, B.	MS
95	Shah, R.	MS
96	Shah, S.T.	PhD
97	Shah, S.	MS
98	Shahani	MS
99	Shea	MS
100	Sheth	PhD
101	Shetty	MS
102	Shih	PhD
103	Shim	PhD
104	Shirazi	MS
105	Sood, S.	MS
106	Sood, V.	MS
107	Srinivasan	MS
108	Srivatsa	MS
109	Suneja	PhD
110	Tavakoli Sabour	PhD
111	Thakker	MS
112	Thomas	PhD
113	Uchil	MS
114	Venkataraman	MS
115	Vesaratchanon	PhD
116	Vijayaraghavan	PhD
117	von Behren	MS
118	Vora	PhD
119	Wnek	PhD
120	Wu, P.	PhD
121	Wu, S.	PhD
122	Xu	PhD
123	Zhang	PhD
124	Zheng	PhD

■ PhD □ MS

FG-9-1: List of my advisees

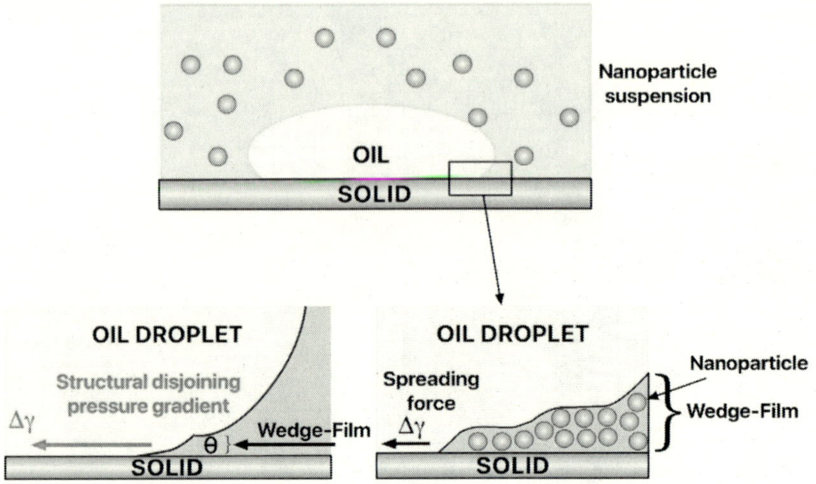

FG-9-2: Nanoparticle structuring in the wedge film resulting in the structural disjoining pressure gradient at the wedge vertex

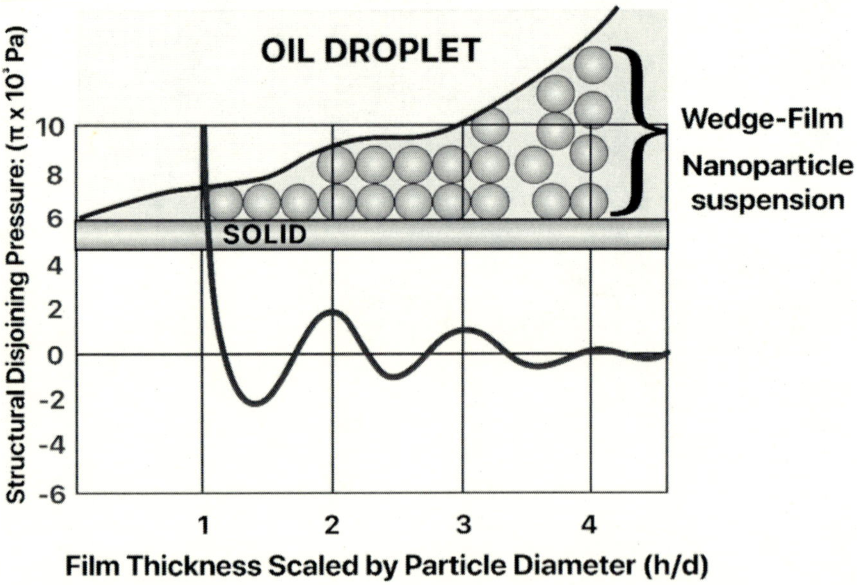

FG-9-3: Structural disjoining pressure on the walls of the wedge

Some Notable Awards and Honors given to Darsh T. Wasan and his Research Teams between 1986–2016

- NSF, 1986: Special Creativity Award, with Dimitri Gidaspow, for research on the separation of particles from nonaqueous media based on electrokinetic phenomena
- NSF, 1988: Special Creativity Award, for research on the discovery of ordered microstructures in thin liquid films
- American Institute of Chemical Engineers, Chicago Section, 1989: Ernest W. Thiele Award, for original and significant contributions to the practice of the chemical engineering profession
- Bulgarian Academy of Sciences, 1990: Asen Zlatarov National Award for Chemical Sciences, for research publications in ordered microstructures in thin liquid films
- American Society of Engineering Education, 1991: 3M Lectureship Award, for outstanding contributions to fundamental chemical engineering theory and practice
- American Chemical Society, 2000: National Award in Colloidal and Surface Chemistry for pioneering theoretical and experimental investigations on the liquid interface in dispersed systems, especially the introduction of a novel colloidal stabilization mechanism for macro- dispersions and for the ingenious translation of fundamental results to full-scale technology
- American Institute of Chemical Engineers, 2002: Thomas Baron Award, for advances in fluid particle interaction forces
- National Academy of Engineering, 2004: Elected Member, for pioneering research, inspirational teaching, and novel technology development in colloidal processing and interfacial rheology
- American Institute of Chemical Engineers, 2005: Alpha Chi Sigma Award, for fundamental and applied research in chemical engineering
- American Institute of Chemical Engineers, 2016: Elsevier Lifetime Achievement Award, for particle technology

William A. Wulf, President, National Academy of Engineering — Darsh Wasan — Craig R. Barrett, Chief Executive Officer, Intel Corporation

My induction into the National Academy of Engineering in 2004. Election to the NAE is considered to be one of the highest professional honors an engineer can attain.

With family, colleagues and former students under the statue of Einstein at NAE in 2004

With Norman Li, who nominated me for induction into the National Academy of Engineering, in 2017

With my research group in 2004

With Jongju Lee at his graduation in 2017

Usha and I with (left to right) Rama Ramakrishnan, Daniel Huang, Nizar Djabbarah, Sanjay Srivatsa, Raju Borwankar, and Alex Nikolov in 1998

With Taylarides at Stevens Distinguished Lectureship, Syracuse University in 1991

With Shekar Shetty in 1989

With Venu Menon at his graduation from IIT in 1986

With my former student Nizar Diabbarah in 2015

With former students Jeff Perl (left) and Carl Pasquarelli (right) in 2018

At my birthday celebration in 2018: (left to right) Gopi Sethumadhavan, Dharam Punwani, Anoop Chengara, me, Ravi Randhava and Angsuman Dutta

With former student Manu Vora in 2012, receiving IIT Alumni Award

Chapter 10
Epilogue

"Come in to learn and go out to serve," are words that have been guiding me in all the steps along the otherwise challenging journey of my life.

I firmly believe that education lifts you and helps you become a more knowledgeable person, ready to meet all the challenges in a world profoundly complex. Service allows you to better utilize your innate talents and capabilities, helping you to create an environment more inviting, beautiful, and lasting. I believe that the customer is the king, so I have treated all my students as kings.

IIT taught me to cultivate the art of the possible; it taught me to chase dreams of my own as a prologue for a rewarding life yet ahead. Thus far, my life has been challenging, enjoyable, and full of hard work. Though luck is always at play in our lives, I consistently have tried, year after year, to do my very best to seize each opportunity that has come across my path in life.

Deep down, I strongly feel that I have accomplished much, because my childhood had prepared me to work hard to be successful. Most importantly, though, I have been blessed by the words of Waheguru Ji, my god, who has been guiding me throughout my journey in life.

The stunning trauma brought on by the partitioning of my homeland and the ensuing hateful religious clashes still loudly echo inside of me. Growing up in a refugee camp in Bombay was very difficult, for we lacked all creature comforts and the basic necessities. When we fled south, my family and I lost more than our land and our earthly possessions.

At times, I truly feel that it was this very painful experience that I needed to become the person I am today. Because of our exile, I was able to attend the Guru Nanak school; after all, every cloud has its silver lining. I will never forget those who lived in the camp, for they worked so very hard to serve others. Many students in my school worked just as hard, but they were not able to substantially change the course of their lives, possibly because they did not have the same opportunities that I had—my brother was already abroad and that is how I came to the United States.

Perhaps, unlike some others, I have been blessed with the ability to know that even minor opportunities could go from possibilities to successful enterprises, such as the creation of the National Center for Food Safety and Technology.

My school's motto, in the opening lines above, left a profound impression on my juvenile mind. The whole academic enterprise of learning, teaching, research, and administration has been engrossing to me; it is all that I have needed to be fulfilled professionally. For me, the focus of the academic mission has always been service; I strongly feel that what I have been doing is not just a job. I have always seen myself providing a service for students and preparing them to succeed in life.

I never experienced the scope of the academic mission as being too grueling or too overwhelming. In my childhood, I experienced many discomforts, and I learned to accept them as the essence of what life was. Subsequently, I was able to do the best with the hand that I was dealt.

With unconditional love, my parents had prepared me to work hard in order to succeed, though they also encouraged me to heed the wisdom of the ages and free myself of what has been called the "tyranny of expectation."

My father's strengths and determination to never give up taught me how to persevere in the face of overwhelming loss and in seemingly impossible odds. My mother's gentle ministrations taught me the importance of lifelong kindness and generosity to others, even when one may not have much to give or to share.

Madan, my oldest brother, was the very embodiment of hard work; he was the man I most aspired to emulate. I remember well how, day after day, he treaded down the well-worn path to the village school, and decades later, the joyful surprise in his eyes on that very first day I set foot in his office at the University of Illinois. Madan was a great inspiration to me during my student years.

The years went by. I truly believe that, later in my career, there was no higher honor than being inducted into the National Academy of Engineering, one of the highest honors bestowed upon an engineer in the United States. Although this award was in recognition of my dedication and hard work of my students and coworkers, it had all been made possible through the blessings of my Waheguru Ji. I am indebted to my good friend, Norman Li, the president of NL Chemical Technology, Inc., for nominating me for this honor.

My numerous travels worldwide to globalize IIT made it possible for me to also reconnect with many of my former students and our alumni. Over a period of 20 years, I had made new friends oversees. But long travels to the Far East took a toll on my health, and the last time I was in Beijing, I suffered a moderate stroke.

My colleagues and friends have often wondered how I have been able to maintain a balance while being very active in teaching, research, and academic and administrative responsibilities, while also devoting full effort to serving my students and others.

I must admit that all these activities often came with sacrifices in my life, and I did not fully appreciate the issue of a work/life balance. One of my deepest regrets is that, at times, my family life appeared to have become second to my career. My work at IIT would often overflow into the most basic parts of family life, an agonizing experience that many others in all professions have also inevitably faced.

While we lived in the faculty apartments on campus, I would often go back to my office after our daily family dinners and work with my research students. I knew well that, instead, I could have spent quality time with my children before they were put to bed. After we moved off campus, I also missed attending many of my sons' games in school and their various extracurricular activities. Still, I am very proud of our two sons, Ajay and Kern, my pillars of joy, who now have successful careers of their own. I am also very glad that they have been able to keep a better work/life balance.

Among several rewards, I have been strongly embraced by a community of those who viewed education as a foundation to helping others. My memories of this rich and diverse community are the many plans that I was involved with for improvements, services, and activities; the countless discussions; and yes, some heated arguments.

From the inspirational Professor Wilke, who literally gave the coat off his back to this impecunious student from India, to the countless insightful discussions and deliberations with my colleagues at IIT, I have indeed been fortunate to work with some amazing individuals with impeccable credentials and stellar accomplishments. There isn't sufficient space here to name them all, although each person has had a unique role in my life. Their dedication and tireless service to students have greatly helped me along my own professional journey and have created new opportunities for my own growth and understanding of the world around me.

The visions and the leadership of Presidents Rettaliata, Martin, Feldberg, Collens, Anderson, and now, Cramb, not only guided our university, they also had a profound effect on me. I learned much from the members of this learned and fraternal community. I am thankful for the opportunity to among all those who have walked through the welcoming doors of IIT.

The countless hours that I spent in the lab, while ironing out the wrinkles in the theses of my students, would not have been possible without the generous help of Usha, my wife. She has been

Waheguru Ji's greatest gift to me. She has been a wonderful wife and mother; she worked tirelessly to raise our children, rarely taking any vacations.

I am immensely grateful to my late parents who arranged our marriage. A quintessential wife and a mother, Usha has unfailingly been there for me, for our children, and for our grandchildren. Her sacrifices, with a comforting, quiet, and modest presence, have kept our family going strong, year after year. They have also made it possible for numerous others who always felt the welcoming comfort and warmth of our home. Her patience and perseverance made it possible for me to fully enjoy my professional dreams, as well as the dreams of my students and of my researchers.

Together, Usha and I have much to treasure, especially the priceless gifts of Arjun and Ashwin, the children of our children. Together, they never fail to remind me of some of the best moments of my life—when our sons were born, when our grandsons were born, and now, when I am a joyful and proud witness to their many accomplishments.

May they not only work hard but also maintain a more fulfilling work/life balance. May they find a profession that they will really enjoy and not pursue a career just for the income. May they remember that not everything in life is determined only by our own efforts, for there will be countless surprises. May they have faith, as well, and aspire for a spiritual life, in the service of others, and with no expectations of the fruits of their labor.

With our golden wedding anniversary now behind us, Usha and I are together looking forward to the rest of our years as we continue remembering the words of an enduring blessing for newlyweds: “May you both grow old on the same pillow.”

This is the final chapter of this compendium of my life thus far. With courage and persistence, I shall continue to look forward to accomplishing more, helping students, and guiding them along the chosen paths while they chase dreams of their own.

I still cannot say if I ever will retire, for I could never tire of the sublime siren call: Come in to learn and go out to serve.

Kern's graduation from Illinois State University in 1995

Usha and I with Ajay at his graduation from the University of Illinois Chicago College of Medicine in 1988.

Ajay's wedding with Simone in 1999

A family portrait in 2010

With grandson Arjun in 2006

With Usha and Arjun in my office in 2009

Arjun's eighth grade graduation in 2018

My 80th birthday celebration in 2018

Madhavan Nayar and his wife at his and my birthday celebrations at my home in 2018

Appendix 1
ILLINOIS INSTITUTE OF TECHNOLOGY (INDIA) PRIVATE LTD.

By D. Rajagopalan

How It All Started

It was 7 a.m. on an early September day in 1996 when my friend Wing Commander S. Venkatesam (retired) called me at my residence in PTP Nagar, Thiruvananthapuram, India, asking me to take a call from a professor at Illinois Institute of Technology in Chicago—a total surprise. Venkatesam, who took voluntary retirement from the Indian Air Force, joined Motorola India Electronics Limited (MIEL) in Bangalore and rose to become managing director of Motorola, China Operations. Even after retirement from the armed services, he was in touch with me.

Nevertheless, I was surprised to hear from him calling from Chicago. He, along with his colleagues, was visiting Professor Darsh T. Wasan, vice president of international affairs at Illinois Institute of Technology.

After exchanging pleasantries, Professor Wasan asked me whether I would meet him sometime in late December 1996 during his visit to Bangalore. He mentioned that he was planning to offer master's degree programs online for working professionals in India, in particular to employees of MIEL at Bangalore. On a visit to Motorola Headquarters in Chicago, Venkatesam and the team called on Professor Wasan seeking help (to be explained later). During the interaction, in response to a query from Professor Wasan about running their operations in India, Venkatesam had suggested a few names, including mine. I was to retire from the Indian Air Force by the last day of December 1996.

Venkatesam, who was known as Venky in the Air Force, was known as Sam at MIEL and in the U.S. Sam, on his visit to Motorola headquarters in Chicago, was accompanied by Shibiraj, director of finance and S. Vidya Sagar, director of human resources at their Bangalore facility. An earlier managing director at MIEL, a person of Indian origin hailing from Pune, India, had requested that Pune Institute of Advanced Technology (PIAT) offer a master's program in software and telecommunication at its facilities in Pune, indicating his willingness to recruit the graduates of such a program to work at MIEL. The institute put together a curriculum and started the program. Quite a few engineers who successfully completed the master's program were recruited by MIEL. Soon, about three years later, a few teachers who taught courses in telecommunications left

PIAT during the middle of the semester. Unable to complete the coursework, PIAT had to postpone the exam and approached the then MIEL Managing Director Roger Fordham to request assistance in resolving the crisis. This was during mid-1996, and Fordham lost no time in sending a team to Motorola headquarters in Chicago, advising the team to seek assistance from Illinois Tech. The team, on the advice of the staff at Motorola headquarters, met Professor Wasan, the vice president for international affairs at Illinois Tech, and asked him to depute a professor from the Illinois Tech campus to complete the curriculum at PIAT, Pune. Venky (Sam) called me from Professor Wasan's residence.

Motorola in the U.S. had established a software development center in Bangalore, India, by 1991. Robert Galvin was the chairman and managing director of Motorola Group and was also the chairman of the policy committee at Illinois Tech. Motorola and Illinois Tech had an excellent relationship since World War II when the professors of Illinois Tech enabled Motorola to build communication equipment for the U.S. Armed Forces. It was during the policy committee meeting held at Illinois Tech sometime during mid-1996 that Galvin suggested the university offer some of its cutting-edge programs in advanced technology to engineers at MIEL too. During that time, engineers at MIEL were working on the Iridium Satellite Project, seeking to provide communication across the globe using handheld phones. It is worth mentioning that the cellular phone also was invented by Marty Cooper, an alumnus of Illinois Tech and an employee of Motorola, U.S.

Most instructors and staff members at Illinois Tech did not have any clue on the solution to the requirement posed by Galvin in the policy committee meeting. Professor Wasan, a migrant from the North-West Frontier Province in pre-independent India, had settled down in Mumbai during India's freedom struggle. He had been in the U.S. since his undergraduate study. He was working on the suggestion from Galvin and started his interactions with a few people from India like Sam Pitroda, a resident of Chicago, to find the ways and means, a solution to the requirement posed by Galvin. Seizing the opportunity during the visit of Venkatesam to Chicago, Professor Wasan traded with the visiting team from MIEL, agreeing to depute a teacher from Illinois Tech to PIAT and, in return, asked MIEL to host a representative of Illinois Tech on their premises to enable Illinois Tech to offer master's degree programs to the engineers of MIEL. Immediately thereafter, Professor Wasan deputed Suresh Borkar from the electrical and computer engineering (ECE) department at Illinois Tech to Pune to complete the unfinished coursework for the students at PIAT. In the next two weeks or so, MIEL agreed to host the representative of Illinois Tech at their premises at Bangalore.

It was a masterstroke by Professor Wasan, while others at Illinois Tech were groping in the dark, which resulted in Indians in India getting a quality American education at an affordable cost without quitting their jobs. All graduates, more than 600 in number to date, owe a lot to Professor Wasan.

Selecting an Illinois Tech India Operation Director

Professor Wasan, during his interaction with the visitors from MIEL, also asked them to identify a person to run their operations in India. In fact, he pointedly asked Sam—the only engineer in the group of three from MIEL that visited him in

September 1996—whether he would take up the assignment. Quoting MIEL rules, Sam declined the offer but agreed to identify a person. The only qualification requirement was that the person chosen should have a master's in engineering, if not a higher degree. Over the next few days, Sam proposed three names to Professor Wasan, giving him a brief preview of their qualifications and background. I, Rajagopalan, was one of the candidates with a master's degree from the Indian Institute of Technology, Delhi. Sam, on the advice from Professor Wasan, asked for my curriculum vitae. After scrutiny of the CV, Professor Wasan asked me to visit Bangalore for an interview during his visit in January 1997.

Professor Wasan, accompanied by Robert Carlson, then chairman of the computer science department at Illinois Tech, visited Bangalore and interviewed me over breakfast in Hotel Oberoi on MG Road. The meeting was informal and we got to know each other. I also got to know their expectations of me. The discussions lasted for about 90 minutes. Professor Wasan advised me to visit MIEL that same evening to meet with Roger Fordham, managing director of MIEL.

After a brief meeting between Fordham and Professor Wasan, I spent about 20 minutes with Fordham in his office at Haudin Road near MG Road. After getting to know my background, Fordham asked me, "How do you expect this program to be successful, and how long do you think it would take for us to make a decision on the continuance of the program or otherwise?" A valid question, indeed, as no other foreign educational institute had entered India and offered engineering programs the way Illinois Tech wanted to do. I answered, "The quality education, if made available at an affordable cost, would be apparent in about three years—then we can make a decision to continue or otherwise." I quickly thought that two years being the duration of the program, the third year was necessary for the graduates to make the desired publicity; no amount of publicity through print or other electronic media could have as much impact as word of mouth. Fordham seemed convinced. Professor Wasan and Carlson met me after the brief meeting, and we bid farewell.

I returned to Thiruvananthapuram, and Professor Wasan and Carlson returned to Chicago. By early February 1997, I was offered the job of overseeing the India operations of Illinois Tech, advising me to join at the MIEL premises in Bangalore. I could not join immediately and sought time until early April 1997.

It was Professor Wasan's passion to have an overseas operation in a developing country that resulted in setting up a center in India. He converted a challenge into an opportunity. The requirement of Galvin to provide advanced education to his engineers at MIEL was converted into an opportunity. Professor Wasan saw the larger picture that could get recognition for Illinois Tech in India and, consequently, larger enrollment in India by offering a quality education to all Indians in India.

I joined in April 1997 and reported for work at the MIEL premises in Bangalore. I took a lot of time to understand the program and course details through the available literature like online graduate bulletins, telephone calls and e-mail interactions with Professor Wasan and his staff in Chicago. Having served in Indian Air Force for over was three decades and not having any marketing experience, I had to find my own way of publicizing the master's program in Bangalore amongst the public and, in particular, to the information technology industries. By the time I got started with the marketing activity in April 1997, Professor Wasan had met and talked with several senior executives of information technology companies, convincing them of the merits of the endeavor. A few executives in Bangalore were willing to sponsor their engineers for the master's program. I started visiting these companies and several other companies and apprised the employees of the benefits of the American master's program, the three key benefits being affordability, flexibility, and international recognition.

Appreciating the remuneration that engineers in the information technology industries in India get, Professor Wasan had fixed the tuition fee at $8,000 U.S. for the entire master's program. This was just half of that on campus in Chicago. One could do the program in a period of two years, which could be extended to a maximum of six years. Unlike students visiting the

U.S. for higher education who are required to take a minimum of nine credit hours of coursework in their first semester, as per visa rules, the students in India could take a minimum of even three credit hours, equivalent to one course. This gave the students enormous flexibility in balancing work and study. The degree, being an American university degree, was the same as in Chicago, and students were eligible for work permit H1B visas to work in the U.S. Such work permits helped information technology companies like MIEL, which was sending its engineers abroad, often to the U.S. on offshore duties. One could notice a dramatic change in the lives of quite a few struggling and aspiring engineers who joined the master's program. They wanted to further their education abroad but could not do so because of family responsibilities. It was a boon from Professor Wasan to such engineers.

Subsequently, in February 1998 and thereafter, a few other U.S. universities attempted to offer their Master of Computer Science programs online in India with limited success.

How the Program Was Offered

Between late 1996 and mid-1997, Professor Wasan visited India three times and had been meeting senior functionaries of the information technology companies, giving presentations on the two program offerings: the Master of Telecommunications and Software Engineering and the Master of Computer Science. An executive at IBM asked Professor Wasan about the combination of telecommunications and software engineering as a program, and Wasan convinced him of the requirement offered by Motorola and other companies. Of the thirteen companies spoken to by Professor Wasan, only five companies, including MIEL, sponsored their engineers. The other four sponsoring companies were Infosys, Wipro, Silicon Automation Systems, and VeriFone, a subsidiary of Hewlett-Packard. At the request of Illinois Tech, Motorola had given a seed grant of

$100,000 U.S. to Professor Wasan, as well as office and classroom space at MIEL, to get the program started in India. Professor Wasan had agreed in return to give partial scholarships to employees of MIEL. He ensured the curriculum was the same as at Illinois Tech in Chicago, so the students and company executives in India could appreciate the quality of Illinois Tech's programs.

Because Internet services in India were poor in those days, Professor Wasan advised the Illinois Tech department concerned to record and send videotapes of the lectures to Bangalore by courier each week. Lectures, recorded during the classes in Chicago, were played a week later between six and nine p.m. at MIEL on scheduled dates with prior notice to participating students. The semester pattern followed was the same as in Chicago, with a week's delay in the conduct of classes, examinations, assignments, and projects. All the assignments, answer books, and projects done by the students were sent to instructors concerned at the Chicago campus by courier, thus ensuring the same evaluation standards as on the Chicago campus. Keeping the promise, Professor Wasan offered a reduced tuition fee of $6,000 U.S. for the complete master's program for employees of MIEL, while others paid $8,000 U.S. Thus he liquidated the seed money given by Motorola over time.

The first course was conducted live by Professor Robert Carlson, then chairman of the computer science department between August 4 and 9, 1997, at MIEL for almost eight hours a day for five days, with copious assignments. Students, all 36 working professionals, met Carlson at the Hotel Oberoi, 300 meters away from MIEL, after class hours to seek clarifications. Carlson, unmindful of the exhaustion after nearly eight hours of lecturing, entertained the students' visits. Professor Carlson gave us the question paper for the examination, advising us to conduct the examination after two weeks, so the students could assimilate the course content. Immediately after this course, the second course was offered via videotape in a regular semester pattern lasting 16 weeks. Students, all working professionals, preferred the latter as this mode gave them adequate time for studies while coping with their work schedules at their companies. Typically a working professional spent nearly fifteen hours of study a week that including six hours of watching lectures and about nine hours of follow up studies at home. As in Chicago, students had to successfully complete a minimum of 10 courses to get the master's degree.

How We Grew

The teaching methodology was different from what students in India had experienced in their undergraduate programs. With just 40 hours of teaching (48 class hours, each of 50 minutes, done at the rate of three class hours a week for about 16 weeks) in a semester of four months, working professionals could cope with work pressures and could apply themselves to their studies. Every course had at least one project that demanded the application of the theory and concepts taught in the class and required problem solving. Students enjoyed this experience and shared the same with their peers and managers. Sandhya Khode, location head at MIEL, Hyderabad, proudly expressed her satisfaction to Professor Wasan that students from MIEL, Hyderabad, undergoing the master's programs at Illinois Tech showed a remarkable maturity in thinking and meaningful participation in meetings.

Such proud students started sharing with their friends, their experiences of their studies and their ability to apply those in the workplace, and then, as we expected, their friends and colleagues began to call us and ask about the program details and how to apply for admission. Quoting such interest among working professionals and sharing success stories of our students in the workplace, we also gave presentations at many companies.

I was also aware that many students, pursuing undergraduate studies in engineering colleges in India, could not afford a visit to the United States for higher education—hence, I started visiting a number of engineering colleges in Bangalore. Many engineers applied for our programs on campus and the India online programs.

At this stage, in June 1999, almost two years after commencing our operations in India, we moved the IIT offices to a rental space in Madivala in southeast Bangalore. I also hired an office assistant, who looked after the office and attended to routine queries. This enabled me to spend the afternoon sessions at the MIEL office.

We had two intakes a year, as in Chicago. Slowly and steadily the strength of students increased from 36 in 1997 to about 200 by 2007. The visibility of our university to the public in India improved considerably, and, consequently, enrollment on our campus grew from 137 Indian students on campus in 1997 to 1,308 in 2007. Though a few staff members at Illinois Tech were unhappy that such a quality program as on the Chicago campus was made available to students in India at half the tuition fee, most others started seeing the merits of this venture and responded favorably to the new courses and programs. Gradually the number of programs offered in India increased to 13 and correspondingly the number of courses to about 60 a semester. This catered for the varied interests among working professionals.

As the number of students in India grew, the number of Indian students at Illinois Tech in Chicago also grew at a considerable pace, as mentioned earlier.

In 2008, the financial crisis hit the American economy so badly that American universities started experiencing a decreased enrollment of students. Illinois Tech was no different. It is interesting to note that in the 2004–05 academic year, Illinois Tech had the distinction of admitting the largest number of Indian students among all U.S. universities for graduate programs. The very next year, Illinois Tech was second in place with combined undergraduate and graduate students admitted in any American university. As the American economy grew under President Barack Obama, by 2012, more Indians sought admission to Illinois Tech.

Government Permission

Indian government rules necessitated a government concurrence before any establishment started its business in India. When we offered the first course in August 1997, we had not obtained the permission. But Professor Wasan was very keen to get started. A reputed attorney in Mumbai, Nishith Desai, engaged by Professor Wasan with assistance from Sam Pitroda, advised us that we could start the program, terming it exploratory (as in our view, no other foreign university was offering any program in India then). Simultaneously in July 1997, we applied for permission from government for operations in India. Our case was pending and unattended to in Ministry of Human Resources, Government of India, in New Delhi. Frequently we called the undersecretary, the junior most bureaucrat in the ministry only to get a response that "there was no precedence to process our case." After similar responses a few times, we requested the ministry give priority to our case. We had cited enough justifications. "Brain drain," which meant a virtual exodus of educated Indians to foreign countries, was being discussed widely all over India then. We pointed out that our programs in India would not only arrest brain drain but could result in "brain gain."

By mid-1997, R. Sethuraman, a senior functionary of a local university in South India, filed a Public Interest Litigation (PIL) in Madras High Court in India, seeking a stay on foreign university programs offered in India. He cited the reason that such establishments offered substandard programs, duping Indians. Apparently, he was referring to a local institute that publicized the award of foreign degrees in computer science for students undertaking its program. A few other agencies also participated in this. Nalini Chidambaram, the wife of former Finance Minister of India P. Chidambaram, took up the case for the institute. The Madras High Court gave a ruling, staying the petition with a rider that foreign universities operating in India should apprise applicants that qualifications so obtained may not be recognized in India.

After a few months of regular interaction, P. C. Bhatt, undersecretary in the Ministry of Human Resources, Government of India, sought a copy of the judgment of Madras High Court. We were provided the same on June 20, 1998. It took a few weeks more for an unfavorable decision from the Ministry of Human Resources. Murali Manohar Joshi, the minister, did not approve our case. Based on my past experience with government machinery, I was confident that the ministry could be requested to reconsider its decision, drawing their attention to relevant points. Professor Wasan cautiously suggested that we go through a consulting firm like KPMG for a fee of $10,000 U.S. In my view, we could not afford its services at that cost; more so, we were not sure of the outcome. As the next step, Professor Wasan suggested that we seek help from the legal wing of MIEL. Quoting my past experience with officials in the ministries in the Indian government (while in Indian Air Force, I had the opportunity of working with such bureaucrats), I suggested to Professor Wasan that I visit Delhi, meet with the officials concerned, and hopefully resolve the issue. During my visit to Delhi, I did call on the legal team of MIEL in Delhi. Their suggestion meant longer time with attendant uncertainty. As planned, I visited Undersecretary

P. C. Bhatt in the ministry and explained the issue very well, requesting a reconsideration of our proposal. I cited several reasons for us to be granted permission to operate in India. The undersecretary advised me to write down a case for reconsideration of our proposal. By the same afternoon, I completed the paper and left it with the undersecretary. After a few days, on August 4, 1998, the government gave us the permission. We are probably the only foreign university to have obtained such a permission.

Having obtained the permission from the Government of India, we worked with Nishith Desai to set up an Indian company to oversee the operations in India. All educational institutes in India are exempt from tax as they register themselves as societies (for the benefit of the community). But Desai advised us to register our Indian establishment as a company that would pay tax. The logic was that the government could become suspicious of a foreign university operating on its own in India without paying tax. Thus, we registered a company Illinois Institute of Technology (India) Private Ltd., a wholly owned subsidiary of Illinois Tech, with the Registrar of Companies on December 29, 1998. This was also my 57th birthday.

Growth of Illinois Institute of Technology (India) Private Ltd.

In earlier days while accompanying Professor Wasan to the companies, I had noted down details of points to emphasize during interactions with executives and engineers at the companies. I noticed a demand for an online management program too. During early 1999, through Professor Wasan, I interacted with Zia Hassan, dean of Stuart School of Business, who was willing to offer an online M.B.A. program in India, the way other engineering programs were offered, the only condition being a minimum of 40 students. I had the brochures and applications ready and met with several executives of local companies. We planned to offer the M.B.A. program at the cost of only $10,000 U.S. A total of 29 people applied. Hassan withdrew the offer, citing an insufficient number of students to have the program started in India as the reason.

While quite a few engineers at MIEL, Bangalore, were pursuing the engineering programs, those in other locations of MIEL, particularly in Hyderabad, showed enormous interest in programs on offer then. So we opened a center in Hyderabad too.

Illinois Tech (India) recruited two persons to assist me in the Bangalore office, one to look after the administration and the other to move around the colleges in Bangalore to publicize our programs and the university. Over the next few years, slowly but steadily, we established four other centers—in Chennai, Pune, Mumbai, and Delhi—assigning the local administration and publicity to a location head at each place. Until about Fall 2012, we saw a steady increase in the number of students and also the number of companies sponsoring their engineers. We persuaded a number of departments at Illinois Tech to offer new programs that resulted in more India online programs.

AICTE

In December 2003, the Government of India came out with a notice that no foreign university could operate in India without permission from the All India Council for Technical Education

(AICTE). We responded immediately with the application, only to be advised that the government was not ready with their procedures to vet our application. Through protracted correspondence with officials at the AICTE and a number of meetings with them, in the year 2007 we obtained a decision. Accreditation could not be granted for the programs offered by our university in India, for two reasons: 1) Technology programs cannot be taught online, and 2) An educational activity cannot be performed by a company (that is designed to make profit). In a later ruling, the Supreme Court of India conveyed that engineering programs offered on online and via correspondence mode are not recognized in India. During the intervening years, many academicians of repute advised us that we need not seek concurrence/accreditation from AICTE, as we do not have brick and mortar educational premises in India and no classes are conducted physically in India. Though it was comforting, we did not give up our dealings with AICTE until 2007. We were interested in advancing the knowledge of instructors in small Indian engineering colleges and thought that completing our master's would have a multiplying effect in raising the teaching standards in these colleges. We did not succeed in this approach because the instructors' promotions were linked to recognition of the master's degree by AICTE. We even offered scholarships to instructors, but it was in vain.

Legal Issues on Tax to Be Paid in India

Initially, all five companies that sponsored their engineers were paying the tuition fee directly to Illinois Tech. We felt that the accounting of the India tuition fee was a cumbersome job for staff at the Illinois Tech campus in Chicago, and hence, we started collecting tuition fees in Indian currency. We had obtained permission from the Reserve Bank of India to collect the tuition fees from students in India in Indian rupees and convert the share of Illinois Tech in U.S. dollars and repatriate them. Things were going smoothly until one company insisted on deducting local taxes and paying the rest to us. This was not consistent with the practice we followed with other companies wherein no tax was deducted. It was our understanding that tax payable to the Government of India would be on the profit we made as Illinois Institute of Technology (India) Private Ltd. We sought a clarification from the government's income tax department.

Instead of answering our question, the department sent a staff member to our premises to study our records. After scrutinizing the relevant papers, the income tax department official felt that we should deduct tax on money received from companies and students before repatriating the share of Illinois Tech to Chicago. We contested this. The matter was referred to the income tax authorities in the department that decided the case in favor of the department after a protracted hearing. We filed a case in the Income Tax Tribunal, challenging the decision of the department. The tribunal heard our case after about six months, and the verdict was in our favor. S. R. Ramasubramanian, a well-known chartered accountant, argued our case for 90 minutes, quoting several rules and judgments from voluminous documents.

The income tax department then went to the Karnataka High Court, contesting the decision of the tribunal. About three years later, the Karnataka High Court heard the case and upheld the decision of the tribunal. Unrelenting, the income tax department filed the case in Supreme Court, and the case has been pending since 2013.

The moment the income tax department made the first claim of taxes, we went back to the earlier practice of Illinois Tech directly collecting tuition fees from companies and students.

Interaction with Staff on Campus

It was in 1999 that I first visited the Illinois Tech campus in Chicago, two years after I joined the university as its representative in India. Professor Wasan was visiting India almost three times a year and was visiting a number of companies and colleges, seeking their partnership in our programs. I got to know more about our campus, programs, and courses during my interactions with Professor Wasan and also during his discussions and presentations at various locations.

Most professors on campus did not evince any interest in the programs offered in India, barring those who were already offering their courses in India. A few instructors were not interested in offering their courses in India as they did not want to evaluate answer books from India a week after they did campus answer books. A few other instructors did not offer their courses, citing a laboratory requirement as a reason. A few even resented the lower tuition fee in India and a few voiced their concerns openly to me during my visit. My answer to them was brief: that the number of Indian students on campus was increasing and hence they should consider money spent in offering programs in India as an investment. Also, I mentioned that Indians cannot afford to pay a higher tuition fee as they are paid less compared to U.S. standards.

Financial Crisis

The Indian establishment of Illinois Institute of Technology (India) Private Ltd. was funded by the Illinois Tech campus in

Chicago, as planned by Professor Wasan. As could be expected out of any business venture (Mr. Syamal Gupta, past director at Tata Sons, India, and also chairman of the advisory committee for our India operation, abhors the use of word "business" while discussing education), we too suffered financial losses and in about six years or so the accumulated losses amounted to about Rs 64 lakhs. In a board meeting held in Nishith Desai's office in Mumbai, the financial situation was discussed, and Desai, a board director, was shocked to learn of the cumulative losses. He raised an alarm that we may have to approach the ministry concerned in the Government of India seeking a concession for continued operation, lest we be declared a sick Industry and be asked to close down. It is to be recalled that the Indian establishment of our university was registered as a tax paying company and not a Section 25 company exempt from paying tax. Professor Wasan was quite concerned and seemed visibly disturbed.

On returning to Bangalore on the evening flight from Mumbai, I consulted my daughter, who was working at KPMG. Promptly, she mentioned that there was no concern as education activity does not come under the purview of the Bureau of Sick Industries in the Government of India. There was a huge sense of relief as she corroborated the statement with related rules. I lost no time conveying the good news to Professor Wasan that same night. I was sure that Professor Wasan, too, heaved a sigh of relief.

In the next few years, we not only managed to wipe out the losses but made a profitand started paying tax to the Government of India. We also started organizing the visits of professors from Illinois Tech's Chicago campus to recruit students for Illinois Tech's on-campus programs.

By 2014, Georgia Institute of Technology, U.S.A., also started offering an online master's program in computer science for half the tuition fee that we levied. The model followed by the institute was very similar to IIT, but included a few riders not easily understood by the students, unless they asked specific questions.

COUNSELING

Mahatma Gandhi, father of India, once stated that "customer is the king." We faithfully followed this and treated all our students as kings. But for them, we would not exist. We paid great attention to their queries and answered all to their satisfaction. We were known for our promptness in responding to students' queries. Word of our credibility spread and more working professionals from companies joined our programs. We made it a point to train all our staff members in various locations well so that our reputation was maintained. At times, we could even intervene in the process at the Chicago campus suggesting corrective measures.

Appendix 2
STARTING THE CHINA OPERATION

By Tim Zhang

In the autumn in 2003, Dr. Wasan visited Beijing and met with Jeffery Li and me. It was obvious that Dr. Wasan had talked about his vision and interest to Jeffrey before regarding increasing IIT's exposure to the Chinese-student community and expanding program exchanges with leading Chinese universities. Jeffrey asked me to meet Dr. Wasan to see whether I couldbe of any help. Dr. Wasan told me about IIT's success in India, and his confidence in IIT's breaking into the Chinese market encouraged me to follow up during my vacation back to the U.S. in December 2003. I was impressed by his passion and enthusiasm.

I drove from Michigan to Chicago in December to meet Dr. Wasan. I proposed a detailed plan on IIT's China operations and sent it to Dr. Wasan by the end of December, and a consulting agreement between IIT and myself was executed in the first of week of January 2004, which meant that I had the mandate to roll out IIT's China operation immediately. I have to say, two factors made this happen. First, Jeffery Li as an IIT alumnus asked me to help, and I would not turn down Jeffery for his love of IIT; second, Dr.

Wasan was so confident in his vision, and his trust in me to get things done made it impossible not to work hard to deliver what he was expecting. The results of IIT's China operation proved that Dr. Wasan was a visionary; he was right about China, and he had a good eye for people and selected the right person to help start this operation. I was encouraged to devote my time and energy to build and lead a team of four to execute the plan I proposed, even though it was obvious that IIT did not have much advantage over competing universities from the U.S., which were better known to Chinese families and universities.

In the first four months, I asked my colleague in my office, Ali Zhang, to help out IIT matters, and in May 2004, I hired Vivien Jin, then a journalist at the college student magazine, the only magazine distributed on Chinese university campuses across China. Vivien had a smile on her face all the time, was very dedicated to work, and had a strong will to pursue excellence at whatever she did. The fact that later Google China's chairman asked her to become a recruiter for Google demonstrated how good she was. Vivien met the Google chairman once at a conference when she accompanied IIT Institute of Design Dean Patrick Whitney, who was a panelist in the same conference. Right after the conference, Google gave Vivien an offer that was hard to turn down. Vivien came to me to ask my permission to join Google; there was no way I could say no to her. Vivien had worked for two years and done a great job for IIT. I was so proud with my recruiting of, first Vivien Jin, and later Hua Qu, who has been dedicated to IIT China operations since then.

IIT's first China Office opened in mid-2004, on the second floor of Guanghua Chang'an Tower, an office overlooking the Chang'an Avenue, the central line dividing north and south Beijing.

To increase awareness of IIT in China, we proposed a series of initiatives that were instrumental. A few key initiatives included the following:

1) We translated and fine-tuned a Chinese introduction to IIT and its selected programs that demonstrated IIT's strength in certain fields to attract like-minded students to apply.
2) We participated in international education shows in selected major cities, including Beijing, Hangzhou, Shenzhen, Shanghai, Guangzhou, etc.
3) We planned and organized the Design for the New China Market conference in Beijing, held December 1–2, 2005, which was such a hit and an influential event and was cohosted by the IIT Institute of Design and the State Intellectual Property Office of China.
4) We conducted interviews on Sohu and Sina to promote IIT programs in China.
5) We reached out to IIT Chinese alumni and set up the China IIT Alumni Club, which has become a community and helps to spread the IIT brand and influences.

From the second year, we started to see an obvious increase in student applications, and we have expanded joint programs with Chinese universities.

Appendix 3
STARTING THE IIT PARIS OFFICE

By Vanita Misquita

The IIT Paris Office was officially established in 1998. Since then, IIT has never looked back. I helped to create a double degree master's program, one of the first in the nation, whereby French students could pursue an IIT master's degree in their fifth year of study and receive two degrees at the same time: the IIT master's degree and their home institution's equivalent degree. The first partnership was, not surprisingly, with my husband Philippe Marc's school, ENSEA (Ecole Nationale Supérieure de l'Électronique et de ses Applications) in Cergy, France. Again, it was uncharted territory also for Philippe and ENSEA. Thanks to Philippe's dedication and conviction that it was a good program, he managed to convince the director (equivalent to the president of a U.S. university) and other faculty members. ENSEA professors, though initially resistant, found that their best student was interested in this program for two reasons: 1) The possibility of receiving an American master's degree and the equivalent degree from his or her home university in one year; and 2) Because of IIT's association with Motorola at the time (through one of the Board of Trustees members). The following year, the number of students from ENSEA alone increased to three, and slowly, other partner schools in France were added through ENSEA's network. After a three-year period, I expanded the partnerships to additional schools in France and in Spain as well—34 in total, 14 in France and 20 in Spain.

Appendix 4
The Darsh T. Wasan Lectureship Series

The Darsh T. Wasan Lectureship Series was established in 2008 in honor of IIT Motorola Chair and Vice President of International Affairs Darsh T. Wasan's extraordinary contributions to Illinois Institute of Technology over the last 50 years. The Wasan lectures seek to bring some of the best and brightest internationally known leaders to speak about cutting-edge topics on global issues.

Ellen Ochoa
Former Director of NASA's Johnson Space Center in Houston, TX

2019 "Building and Operating the International Space Station"

Ellen Ochoa was the director of NASA's Johnson Space Center in Houston, Texas, from 2013 until her retirement in May 2018. She became the first Hispanic woman to go to space when she flew on a nine-day mission aboard the shuttle Discovery in 1993. Ochoa has flown in space four times, logging nearly 1,000 hours. She currently serves on several boards, including as vice chair of the National Science Board; director for Service Corporation International, Mutual America, and the Gordon and Betty Moore Foundation; and chair of the Nomination Evaluation Committee for the National Medal of Technology and Innovation.

Prior to her career as an astronaut, Ochoa was a research engineer and holds three patents for optical systems. She received a B.S. in physics from San Diego State University and both an M.S. and Ph.D. in electrical engineering from Stanford University. She is honored to have six schools named for her. She has been inducted into the Astronaut Hall of Fame, the California Hall of Fame, and the International Air & Space Hall of Fame. She is a fellow of the American Association for the Advancement of Science, the American Institute of Aeronautics and Astronautics, the National Academy of Inventors, and the Optical Society of America. She is the recipient of honorary doctorates from the University of Pennsylvania, the Johns Hopkins University, and Worcester Polytechnic Institute and is the Presidential Distinguished Rank of the Senior Executive Service.

Rohit Prasad
Vice President and Head Scientist, Amazon Alexa

2018 "The Story of Alexa: AI for Daily Convenience"

Rohit Prasad is vice president and head scientist of Amazon Alexa, the voice service that powers Amazon's family of Echo products, Amazon Fire TV, and third-party offerings. Prasad leads Alexa research and development in artificial intelligence technologies aimed at making interaction with Alexa a magical experience for customers. Prior to Amazon, Prasad was deputy manager and senior director of the speech, language, and multimedia business unit at Raytheon BBN Technologies. In that role, he directed U.S. government-sponsored research and development initiatives in speech-to-speech translation, psychological health analytics, document image translation, and STEM learning. Prasad is a named author on more than 100 scientific articles and holds several patents. He earned his master's degree in electrical engineering at Illinois Tech and a bachelor's degree in electronics and communications engineering from Birla Institute of Technology in Mesra, India.

John P. Holdren
Science Advisor and Director of the White House Office of Science and Technology Policy under U.S. President Barack Obama

2017 "Global Climate Change and U.S. Interests: What We Know. What We expect. What Should We Do."

John P. Holdren was President Obama's science advisor and the senate-confirmed director of the White House Office of Science and Technology Policy (OSTP) from January 2009 to January 2017. He was the longest-serving science advisor to the President in the history of the position dating back to World War II.

Holdren is a member of the U.S. National Academy of Sciences, the U.S. National Academy of Engineering, the American Philosophical Society, the American Academy of Arts and Sciences, and the Council on Foreign Relations. He was one of the first winners of a MacArthur Foundation Prize Fellowship (1981) and is a former president of the American Association for the Advancement of Science (2006–07).

Prior to joining the Obama administration, Holdren was a professor in both the Kennedy School of Government and the Department of Earth and Planetary Sciences at Harvard University, as well as CEO of the independent, nonprofit Woods Hole Research Center, where he now serves as senior advisor to the director. From 1973 to 1996, he was on the faculty of the University of California, Berkeley, where he cofounded and co-led the interdisciplinary graduate-degree program in energy and resources. He earned S.B. and S.M. degrees from the Massachusetts Institute of Technology and a Ph.D. from Stanford University in aerospace engineering and theoretical plasma physics.

Holdren is currently the Teresa & John Heinz Professor of Environmental Policy, Harvard Kennedy School and Professor of Environmental Science & Policy, Department of Earth & Planetary Sciences at Harvard University

Sam Pitroda
Chairman of the Pitroda Group

2016 "Technology, Innovation, Entrepreneurship, and Social Change"

Sam Pitroda is an internationally respected development thinker, policy maker, telecom inventor, and entrepreneur, who has spent more than 50 years in information and communication technology and related global and national developments. After receiving an M.S. in physics in India, Pitroda came to the United States in 1964 to study at Illinois Institute of Technology. After graduating with an M.S. in electrical engineering in 1966, he worked at GTE and in 1974, founded Wescom Switching, Inc. When Rockwell acquired Wescom in 1980, he became a vice president at Rockwell. Later, he founded MTI Electronics in Milwaukee and Martek, Ionics and Emnet in Chicago and then returned to India.

Credited with laying the foundation of India's technology and telecommunications revolution in the 1980s, Pitroda has been a leading campaigner to help bridge the global digital divide, focusing on access to technology as the key to social change. During his tenure as advisor to Prime Minister Rajiv Gandhi, Pitroda headed six technology missions related to telecommunications, water, immunization, literacy, oil seeds, and dairy. He was also the founding chairman of India's Telecom Commission.

In the mid-1990s, Pitroda founded WorldTel in London with the International Telecom Union (ITU) to help finance privatization of telecom in emerging markets. He invented mobile wallet in 1996 and set up C-SAM Inc., which was acquired by MasterCard in 2013.

From 2005 to 2009, Pitroda chaired India's National Knowledge Commission and provided a blueprint for reform of the knowledge-related institutions and infrastructure. From 2009, he was advisor to the Prime Minister Manmohan Singh of India with the rank of a cabinet minister on Public Information Infrastructures and Innovations. He was also chairman of the National Innovation Council, Smart Grid Task Force, Railway Modernization, and Public Broadcast Reforms.

At present, he is founder, investor, and chairman of six startups and several not- for-profit foundations including Transdisciplinary University, People for Global Transformation, the Global Knowledge Initiative, and the India FoodBanking Network. He is also a founding member of the UN Broadband Commission and chairman of the m-powering initiative at the ITU.

Pitroda holds more than 100 patents and has published and lectured extensively in U.S., Europe, India, and Latin America. He is a member of the Indian National Academy of Engineering and a foreign member of the U.S. National Academy of Engineering. His awards and recognition include a 2006 Innovation Award from *The Economist*, the 2007 Institute of Electrical and Electronics Engineers' Communications Society Award for exceptional contributions in developing indigenous systems and telecommunications infrastructure in India, and honorary doctorates from universities in the U.S. and abroad. His biography was published in 1992 and his autobiography, *Dreaming Big*, was published by Penguin in 2015.

Muhammad Yunus
Chairman of the Yunus Center, Bangladesh

2015 "Redesign Economics to Redesign the World"

Nobel Laureate Professor Muhammad Yunus, chairman of the Yunus Center, is the father of social business and microcredit, the founder of Grameen Bank and of more than 50 other companies in Bangladesh. In March 2012, *Fortune* magazine named Professor Yunus as "one of the greatest entrepreneurs of our time" for his constant innovation and enterprise.

In 2006, Yunus and Grameen Bank were jointly awarded the Nobel Peace Prize. He is one of only seven individuals to have received the Nobel Peace Prize. He has received 112 awards from 26 countries including state honors from 10 countries. He has received the U.S. Presidential Medal of Freedom and the U.S. Congressional Gold Medal, and he is the recipient of more than 54 honorary degrees from universities across 20 countries. Other notable awards include the Ramon Magsaysay Award (1984), the World Food Prize (1998), the Prince of Austria's Award for Concord (1998), the Sydney Peace Prize (1998), and the Seoul Peace Prize (2006). Yunus was chosen by the Wharton School of Business as one of the 25 most influential business persons of the past 25 years. *AsiaWeek* (Hong Kong) selected him as one of 20 great Asians (1975–95). Ananda Bazaar Patrika (India) selected Yunus as one of 10 great Bengalis of the century (1900–99). In 2006, *Time* magazine listed Yunus as one the top 12 business leaders in "60 Years of Asian
Heroes." In an open online poll, Yunus was voted the second topmost intellectual person in the world on the list of top 100 public intellectuals (2008) by *Prospect* magazine (U.K.) and *Foreign Policy* magazine (U.S.). The *New Statesman* (U.K.) listed him as one of the World's 50 Most Influential Figures (2010). Yunus has appeared on *The Oprah Winfrey Show*, *The Daily Show* with Jon Stewart, *The Colbert Report*, *Real Time with Bill Maher*, *HARDtalk* on BBC, and *The Simpsons*. He has appeared on the cover of *Time* magazine, *Newsweek* and *Forbes* magazine.

Robert S. Langer
David H. Koch Institute Professor, Massachusetts Institute of Technology

2014 "Biomaterials and Biotechnology: Controlled Drug Delivery Systems and Tissue Engineering"

Robert S. Langer, the MIT David H. Koch Institute Professor, is the most cited engineer in history. Langer has written over 1,250 articles, and he has nearly 1,050 patents worldwide. His patents have been licensed or sublicensed to over 250 pharmaceutical, chemical, biotechnology, and medical device companies.

Langer has received over 220 major awards. He is one of seven individuals to have received both the U.S. National Medal of Science (2006) and the U.S. National Medal of Technology and Innovation (2011). He is the recipient of the Charles Stark Draper Prize of the National Academy of Engineering (2002); the Millennium Prize (2008), the world's largest technology prize; the Priestley Medal (2012), the highest award of the American Chemical Society; the Wolf Prize in Chemistry (2013); and the Breakthrough Prize in Life Sciences (2014). He is also the only engineer to receive the Gairdner Foundation International Award. His additional awardsinclude the Dickson Prize for Science (2002); the Heinz Award for Technology, the Economy, and Employment Award (2003); the Harvey Prize (2003); the John Fritz Award (2003), given previously to inventors such as Thomas Edison and Orville Wright; the General Motors Kettering Prize for Cancer Research (2004); the Dan David Prize in Materials Science (2005); the Albany Medical Center Prize in Medicine and Biomedical Research (2005), the largest prize in the U.S. for medical research; the Max Planck Research Award (2008); the Prince of Asturias Awardfor Technical and Scientific Research (2008); the Warren Alpert Foundation Prize (2011); and the Terumo International Prize (2012). He received the Lemelson-MIT Prize (1998), the world's largest prize for invention for being "one of history's most prolific inventors in medicine." In 1989, Langer was elected to the Institute of Medicine of the National Academy of Sciences; in 1992, he was elected to both the National Academy of Engineering and to the National Academy of Sciences; in 2006, he was inducted into the National Inventors Hall of Fame; and in 2012, he was elected to the National Academy of Inventors.

Forbes magazine (1999) and *BioWorld* (1990) have named Langer as one of the 25 most important individuals in biotechnology in the world. In 2002, *Forbes* selected Langer as one of the fifteen innovators worldwide who will reinvent our future. In 2001, *Time* magazine and CNN named Langer one of the 100 most

	important people in America and one of the eighteen top people in science or medicine in America (America's Best). In 2004, *Parade* magazine selected Langer as one of six heroes whose research may save your life. Langer has received honorary doctorates from universities in the U.S. and abroad. He received his bachelor's degree in chemical engineering from Cornell University in 1970 and his Sc.D. in chemical engineering from the Massachusetts Institute of Technology in 1974.

Rajendra Kumar Pachauri Director General of the Energy and Resources Institute, New Delhi, India

2013 "Harnessing Knowledge to Meet the Challenge of Climate Change"

Rajendra Kumar Pachauri is the chair of the Nobel Peace Prize-winning Intergovernmental Panel on Climate Change (IPCC), the scientific intergovernmental body that provides decision makers and the public with an objective source of information about climate change. He is also director general of The Energy and Resources Institute (TERI), a major independent research organization that provides knowledge on energy, environment, forestry, biotechnology, and the conversation of natural resources. Pachauri is a prominent researcher on environmental subjects, recognized internationally for his efforts to build up and disseminate greater knowledge about manmade climate change and to lay the foundation for the measures that are needed to counteract such change. He has been senior adviser to the Yale Climate and Energy Institute (YCEI) since July 2012 and was the founding director of YCEI from July 2009 to June 2012. He is active in several international forums dealing with the subject of climate change and its policy dimensions.

He was awarded the second-highest civilian award in India, the Padma Vibhushan, in January 2008 by the President of India and received the Officer de la Legion d'Honneur from the government of France in 2006. In 2009, he received an Honorary Doctor of Engineering from Illinois Institute of Technology. He has been conferred with the Order of the Rising Sun, Gold and Silver Star by His Majesty Akihito, Emperor of Japan; the Commander of the Order of the White Rose of Finland by the Prime Minister of Finland; the Commander of the Order of Leopold II by the King of the Belgians; and the Mexican Order of the Aztec Eagle by the President of Mexico in June 2012.

Subra Suresh Director, National Science Foundation

2012 "Challenges and Opportunities for Global Science"

A distinguished engineer and professor, Subra Suresh was nominated by President Barack Obama in 2010 and unanimously confirmed by the U.S. Senate to serve as the director of the National Science Foundation.

Prior to his appointment to lead the $7-billion federal agency, Suresh was dean of the School of Engineering and the Vannevar Bush Professor of Engineering at the Massachusetts Institute of Technology. His experimental and modeling work on the mechanical properties of structural and functional materials, innovations in materials design and characterization, and discoveries of possible connections between cellular nanomechanical processes and human disease states haveshaped new fields in the fertile intersections of traditional disciplines. He has co-authored more than 240 journal articles, registered 22 patents, and written three widely used materials science books. In his leadership roles at MIT, Suresh helped create new state-of-the-art laboratories, the MIT Transportation Initiative, and the Center for Computational Engineering; led MIT's efforts in establishing the Singapore-MIT Alliance for Research and Technology (SMART) Center; and oversaw the recruitment of a record number of women faculty in engineering.

Suresh has been elected to 10 different science or engineering academies including the U.S. National Academy of Sciences, the U.S. National Academy of Engineering, and the American Academy of Arts and Sciences. In addition, he has been elected a fellow or honorary member of all the major materials research societies in the U.S. and India and has been awarded seven honorary doctorate degrees. In 2006, *Technology Review* magazine selected Suresh as a top-ten researcher whose research "will have a significant impact on business, medicine, or culture." His many other honors include the Acta Materialia Gold Medal (2006), the European Materials Medal (2007), the Eringen Medal of the Society of Engineering Science (2008), the General President's Gold Medal from the Indian National Science Congress (2011), the Nadai Medal from the American Society of

	Mechanical Engineers (2011), the R. F. Mehl Award from the Minerals, Metals & Materials Society (2012), and the Timoshenko Medal from ASME (2012). In 2011, he was awarded the Padma Shri, one of the highest civilian honors, for his work in science and engineering by the President of India. Since joining the NSF, he has established several new initiatives including INSPIRE (Integrative NSF Support Promoting Interdisciplinary Research and Education), PEER (Partnerships for Enhanced Engagement in Research) in collaboration with USAID, Science Across Virtual Institutes (SAVI), the NSF Career-Life Balance Initiative, and the NSF Innovation Corps. In May 2012, he led the NSF's inaugural summit of heads of science funding agencies from 47 developed and developing countries, which led to the creation of the Global Research Council, a virtual entity aimed at harmonizing global science.

Alice P. Gast
President, Lehigh University

2011 "International Education Today"

Alice P. Gast, a renowned leader in higher education and a distinguished researcher in the study of surface and interfacial phenomena, was appointed the 13th president of Lehigh University in 2006. She is co-author of *Physical Chemistry of Surfaces*, a classic textbook on colloid and surface phenomena, and has presented named lectures at several of the nation's leading research institutions.

Before joining Lehigh, Gast served as the vice president for research and associate provost at the Massachusetts Institute of Technology and held the Robert T. Haslam Chair in Chemical Engineering. She previously spent 16 years as a professor of chemical engineering at Stanford University and the Stanford Synchrotron Radiation Laboratory.

Gast was named one of the top 100 "Modern Era" engineers in the country under the category of "Leadership" by the American Institute of Chemical Engineers. In 2010, she was named to the prestigious post of science envoy by U.S. Secretary of State Hillary Rodham Clinton and the U.S. State Department. Also in 2010, she was awarded an honorary degree from the University of Western Ontario, where she was described as "a scholar, a researcher, and a leader who is a great inspiration to female and male students, alike, who aspire to enter the profession of engineering." Additionally, Gast is a fellow of the American Academy of Arts and Sciences and a member of the National Academy of Engineering.

Gast earned her bachelor's degree in chemical engineering from the University of Southern California in 1980 and her master's and Ph.D. in chemical engineering from Princeton University in 1981 and 1984, respectively.

Susan Solomon
Senior Scientist, National Oceanic and Atmospheric Administration

2010 "A Tale for Our Times: Something for Everyone about Climate Change and the Reasons for Climate Gridlock"

Susan Solomon is a senior scientist at the National Oceanic and Atmospheric Administration in Boulder, Colorado. She received her bachelor's degree in chemistry from IIT in 1977 and her Ph.D. from the University of California, Berkley, in 1981.

She is widely recognized as one of the leaders in the field of atmospheric science. She is well known for having pioneered the theory explaining why the ozone hole occurs in Antarctica, as well as having obtained some of the first chemical measurements that helped to establish the chlorofluorocarbons as its cause.

She received the National Medal of Science, which is the highest scientific honor in the U.S. She also has received the Grande Medaille, the highest award of the French Academy of Sciences, and she is a recipient of the prestigious Blue Planet Prize in Japan and the Volvo Environment Prize. A glacier in Antarctica has been named in her honor: Solomon Glacier. Along with Al Gore Jr., she shared the 2007 Nobel Peace Prize in her role as cochair of the intergovernmental panel on climate change. In 2008, *Time* magazine named Solomon as one of the 100 most influential people in the world. In 2009, she was inducted into the National Women's Hall of Fame. She is a member of the U.S. Academy of Sciences, the Royal Society, and European Academy of Sciences. She is a recipient of numerous honorary doctorates from universities in the U.S. and abroad.

Norman N. Li
President, NL Chemical Technology, Inc.

2009 "Membrane Technology for a Thirsty Century from an Industrial Perspective of Energy, Sustainability, and Globalization"

Norman N. Li is president of NL Chemical Technology, Inc., in Mount Prospect, Illinois. He has a distinguished career of 40 years in the American chemical and petroleum industries. He joined Exxon Research and Engineering Co. after receiving his Ph.D. in chemical engineering from Stevens Institute of Technology in 1963. He also served as a consultant for the U.S. Apollo Moon Landing Project. He joined the UOP Co. in 1982 as its research director of Separations Science and Technology and later became the director of research at AlliedSignal Co. (now Honeywell). In 1995, recognizing the critical needs of clean water in many parts of the world, he founded the NL Chemical Technology, Inc. to develop cost-effective advanced membrane technologies for water purification and desalination.

Li has made major technical contributions to the advancement of separation science and technology, in particular to membrane science and technology. He invented the "liquid membrane" and "mixed matrix membrane" for effective separation and purification of gas and liquid phases. He also invented an energy- saving reverse osmosis membrane for desalination applications. Dr. Li has made significant contributions to the worldwide implementation of industrial separation processes, which include membrane processes, adsorption processes, and for the manufacture of ozone-friendly fluorocarbons.

Li has more than 100 technical publications, 44 U.S. patents and 13 books edited. He has given invited lectures at 80 universities and industrial laboratories around the world.

He is a member of the National Academy of Engineering of U.S., the Chinese Academy of Science, and Academia Sinica. He is also the recipient of numerous engineering awards.

David A. Edwards
Gordon McKay Professor of the Practice of Biomedical Engineering, School of Engineering and Applied Sciences, Harvard University

2008 "Innovation in Global Healthcare, Education and the Arts"

David A. Edwards is the Gordon McKay Professor of the Practice of Biomedical Engineering at the School of Engineering and Applied Sciences at Harvard University. He is a member of the National Academy of Engineering, considered one of the highest honors bestowed upon an engineer in the United States; a three-time recipient of the Ebert Prize of the American Pharmaceutical Association; and winner of various national and international awards, including the AICHE Professional Progress Award for Outstanding Progress in Chemical Engineering. He is the founder of Advanced Inhalation Research (AIR) and also founded *Le Laboratoire*, a new art and science cultural center in Paris's first arrondissement. In addition, he was elected as *Chevalier* of the Order of Arts and Literature of France in August 2008, which is a significant recognition of his contributions to arts, literature, and the propagation of these fields.

With David Edwards in 2008

With my former students and friends at the 2008 inaugural Wasan lecture

With Susan Solomon (right) and Alan Cramb in 2010

With Alice Gast (second from right) and President Anderson in 2011

With my former research students and colleagues at the 2014 Wasan lecture

With Sam Pitroda (second from left) and Muhammad Yunus (center) in 2015

With John Holdren (center) and President Cramb in 2017

The Darsh T. Wasan Lectureship Series is made possible by donations from former students of Professor Wasan and his friends and fellow faculty members.

- Rajev Agarwal (Ph.D. ChE '02)
- Stephen J. and Jacqueline L. Anderson (PSYC '05)
- Hamid Arastoopour (M.S. GE '75, Ph.D. '78)
- Dibakar Bhattacharyya (Ph.D. CE '66)
- Christine Borwankar (ChE '84, M.S. '86) and
- Rajendra P. Borwankar (Ph.D. ChE '84)
- Paul R. Brauer (M.S. ChE '81)
- Man-Sheung Chan (M.S. ChE '78)
- Jaideep Chatterjee (M.S. ChE '90, Ph.D. '96)
- Anoop Dutta (Ph.D. ChE '03)
- Chang W. Chi (Ph.D. ChE '68)
- S .R. Cho* (M.A.S. ChE '66)
- Atul C. Choskey (ChE '73)
- Mary Dawson
- Nizar F. Djabbarah (M.S. ChE '73, Ph.D. ChE '78)
- Gerard J. D'Souza (M.S. ChE '67, M.S. IE '73)
- Angsuman Dutta (M.S. CS '02, M.S. ChE '02)
- David A. Edwards (Ph.D. ChE '87)
- K. M. Gala (M.S. IE '66)
- S. K. Gala
- Alice P. Gast
- Dr. and Mrs. Shailesh B. Godambe (M.S. IE '70, Ph.D. '75)
- Jamshyd N. Godrej* (MAE '72)
- Raghubir P. Gupta (Ph.D. ChE '90)
- Alexander E. Hergan (EE '70)
- Daniel Huang (Ph.D. ChE '85)
- Arthur B. Israel (M.S. ChE '68)
- Wook Jang (Ph.D. ChE '04)
- Joseph Kaellis (Ph.D. ChE '70)
- Richard L. Kao (M.S. ChE '68, Ph.D. Che '88)

- Kalman Koczo
- Kiriprakash Kondiparty (Ph.D. CHE '12)
- Young Sun Kong (Ph.D. CHE '09)
- Dhongchai Lamsam (M.S. IE '70)
- Norman and Jane Li
- Sangwook Lim (Ph.D. ChE '17)
- Steven E. Lindee
- Natalie and Henry R. Linden (Ph.D. ChE '52)
- Richard R. Lindsey (ChE '76)
- Lloyd A. Lobo (M.S. ChE '87, M.A.S. '87, Ph.D. '91)
- Anil K. Malhotra (Ph.D. ChE '84)
- Hansraj C. Maru (M.S. ChE '70, Ph.D. '75) and Usha H. Maru (M.S. ENVE '74)
- Edward H. Mayer
- Venugopal B. Menon (Ph.D. ChE '86)
- William S. Mondschein (ChE '67, M.S. '68)
- R. Nagarajan (Ph.D. ChE '87) Sucha Nand
- Ann Nathan and Walter Nathan* (B.S. ME '44)
- Madhavan K. Nayar (M.S. IE '68)*
- Alexander D. Nikolov
- Charles A. Padgitt (ChE '50)
- Carl H. Pasquarelli (M.S. ChE '80)
- Jeffery P. Perl (ChE '77, M.S. '79, Ph.D. '84)
- Daniel J. Pintar (B.S. ChE '86)
- Jai Prakash
- Gunupati V. Prasad (ChE '82)
- Dharamvir Punwani (M.S. ChE '67)
- Raj Rajaram (J.D. '91)
- T. S. Ramakrishnan (Ph.D. ChE '85)
- Madhav B. Ranade (M.S. ChE '68, Ph.D. '74) and Mary S. Ranade (M.S. Math '67, Ph.D. '74)
- Sarabjit S. Randhava (MS ChE '66, Ph.D. '70)
- Surjit Randhava
- Narayanan Sankar (M.S. ChE '67)

- Narayanan Sankaran (M.S. ChE '75)
- Rolf R. Schroeder (ChE '56, Ph.D. '64)
- Gopi Sethu (Ph.D. CHE '00)
- Syed Muzafar Shah (Biol '74, M.S. '77, M.S. ChE '77)
- Chandrashekar S. Shetty (M.S. ChE '89)
- Jiyoung Shim (Ph.D. ChE '17)
- Ronnie Jo Sokol
- Kehar S. Soodan
- Narayanaswami Srinivasan (M.S. ChE '67)
- Sanjay R. Srivatsa (M.S. ChE '79)
- Ajva Taulananda (Ph.D. IE '73)
- Li Ting (Ph.D. ChE '84)
- Bipin Vora
- Manu K. Vora (M.S. ChE '70, Ph.D. '75)
- Darsh T. Wasan
- Raymond W. Winkler (ChE '45)
- Ralph E. Wozniak (B.S. ChE '68)
- Sho-Chen Wu (Ph.D. ChE '06)
- Wen Xu (Ph.D. ChE '97)
- Hua Zhang (Ph.D. ChE '16)
- Shuming Zheng (Ph.D. ChE '89)

* Trustee, Illinois Institute of Technology

ABOUT THE AUTHOR

Darsh Wasan is an outstanding engineering educator, researcher, and academic leader who came to the United States from India as an undergraduate in 1957 to study chemical engineering, and earned a B.S. from the University of Illinois at Urbana-Champaign, and a Ph.D. From the University of California at Berkeley. Since joining the Illinois Tech Faculty in 1964, he has held nearly every leadership post, including department chair, interim dean of engineering (twice), vice president of research and technology, provost, and senior vice president for academic affairs and, since 1996 has served as the vice president for international affairs. As Illinois Tech's chief international ambassador, he established offices in India, China, Europe, and South Korea, forged alliances with more than 100 universities around the globe, and dramatically increased international student enrollment, achieving the highest percentage among the Association of Independent Technological Universities.

As an educator and researcher, he has supervised and mentored more than 120 graduate students, including 65 doctoral students, and has won excellence in teaching awards from both from his university and the American Society for Engineering Education. He has lectured worldwide and set up research collaborations with faculty abroad, especially in Eastern Europe. He has published more than 400 papers and a textbook, and he holds six patents. Dr. Wasan has been honored with numerous awards for his research accomplishments from several professional societies including the American Institute of Chemical Engineers and American Chemical Society and is an elected member of the US National Academy of Engineering and a fellow of the Indian National Academy of Engineering.

Dr. Wasan served as editor-in-chief of a premier international publication the Journal of Colloid and Interface Science for more than 20 years. To honor him, the publisher named its award for scientists who have made outstanding lifetime contributions to the field the JCIS Darsh Wasan Award.

Acting as an entrepreneur within the university, he secured more than $100 million in funding for Illinois Tech during his tenure and created key initiatives that expanded the university's research capabilities as well as its educational mission. After securing a gift of a five-building suburban campus from industry in 1988, he won funding from the US FDA and founded an innovative national center for research and education in food safety and technology (now called the Institute for Food Safety and Health). In 1997, he broadened the reach of Illinois Tech's online learning, by establishing Illinois Institute of Technology (India) Private Ltd., which has enabled more than 500 working professionals in India to earn cutting- edge master's degrees without leaving home, and has increased awareness in India of Chicago's Illinois Tech

Made in United States
North Haven, CT
10 November 2026

22096546R00150